AF884301

Learn Critical Aspects

of

Pattern & Mould Making in Foundry

By

SHEOJEE PRASAD

Illustrated and Designed by Sheojee Prasad

Rev 3.0 dated 24-Mar-2019

Copyright © 2019 by ReeSaa Pty Ltd, Email reesaa@indiavivid.com

Image of Iron Foundry in Zhenhai by Jim Bowen / CC BY 2.0 has been used to create the book cover

All rights reserved. Without limiting the rights under copyright reserved above, no part of this publication may be reproduced, stored in or introduced into a retrieval system, or transmitted, in any form, or by any means (electronic, mechanical, photocopying, recording, or otherwise) without the prior written permission of the copyright owner

About the Author

Sheojee Prasad is an Author and practitioner with 40 years of experience in Tata Refractories. As a passionate educator and a professional he offers his experience to educate next generation of career professional in the field of refractories, pattern making and mould making.

In this book, Sheojee Prasad makes the task of pattern and mould making in foundry simple.

To begin with, this book explains the concept of pattern and mould making with practical examples and guidelines.

Also by Sheojee Prasad

Pattern & Mould Making in Foundry - Vol.2 (Practical Exercises) by Amazon.com

https://amzn.to/2U3i0yq

Engineering Drawing – A Practical Approach

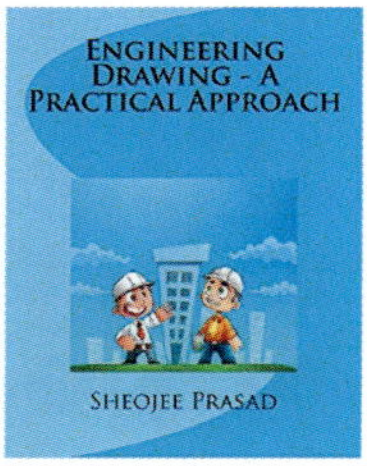

https://amzn.to/2YB3rRh

Mould Design Using Hydraulic Press: For Refractory Bricks – The Practical Way of Designing Moulds

https://www.amazon.com/dp/1542573726

Making Mould With Wood

https://www.amazon.com/dp/1717530265

Jigs, Fixtures and Measuring Instruments

https://www.amazon.com/dp/1797930362

Acknowledgements

I have working experience of about 40 years across Pattern design, Pattern making, Mould design and Mould manufacturing for production of refractory shapes. I had opportunity to manage pattern shop and design wooden & steel moulds during my tenure at Tata Refractories Limited (currently TRL Korasaki Refractories Ltd.)

This book's leverages my experience in Pattern Shop, Wooden Mould shop, Steel Mould shop, Machine shop, Flow Control Mould shop, Mould design and Refractories lining design.

I hope this book will guide pattern shop and foundry personnel as well as those involved or getting into the profession of pattern making & moulding.

I thank my son for inspiring me to document my experiences in the form of this book.

Table of Contents

Machining allowance
Shake allowance
Distortion allowance
Fillets and radius

Chapter – 3 – Nomenclature of Pattern According to Construction Materials

Category of pattern considering construction materials
Plaster of perish pattern
Araldite pattern
Thermocole Pattern
Plastic pattern
Wax pattern
Wooden pattern
Metal pattern
Machines required for making metal patterns
Hand tools and accessories

Chapter – 4 – Category of Wooden Patterns

Category of patterns considering design, construction and use
Solid pattern
Split pattern
Cope and drag pattern
Mounted pattern
Gated pattern
Self core pattern
Follow board pattern
Match plate pattern
Loose piece pattern
Skeleton pattern

Sweep pattern

Chapter- 5 – Pattern Construction Decisions & Step of Pattern Making

Decision on pattern construction
Steps of pattern making
Layout
Contraction scale

Chapter – 6 – Elements of Gating System, Types of Gates, Feeders & Design Considerations

Elements of gating system
Pouring basin
Sprue
Sprue well
Runner
Runner well
Gate/ Ingate
Ingate choke
Types of gating system
Horizontal gate
Vertical gate
Top gating
Bottom gating
Side gating
Gating system design
Turbulence
Bad effect of turbulence
Vent holes
Riser

Solidification shrinkage of metal
Types of risers
Advantage of top riser
Advantage of blind riser
Design consideration
V/A ratio
Solidification
Progressive solidification
Directional solidification
Examples of progressive and directional solidification

Chapter – 7 – Product Design Consideration

Chapter – 8 – Use of Wood, Measurement and Seasoning

Wood
Measurement
Commercial use
Suitable wood for making pattern and core boxes
Seasoning process
Associated materials for making wooden pattern

Chapter – 9 – Wood Working Skill

General operational knowledge for making pattern with wood
Process of joining wooden planks
Typical wood working joints

Chapter – 10 – Brief Introduction of Associated Skill of Pattern Making

Chapter - 1

Introduction to Foundry Technology,

Pattern and Core Boxes

Foundry Technology:

Foundry Technology refers to the manufacturing process that deals with heating metals to a liquid state in furnace and pouring to a sand mould. The object taken out from mould is called casting. The complex objects that are not economical and difficult to make by other processes of forming are made by casting. The most common metals that are cast to a desired shape are

- Aluminium
- Cast Iron
- Steel
- Copper
- Bronze
- Brass
- Magnesium
- Zinc.

Furnaces used in foundry:

- Blast furnaces
- Cupola furnaces
- Electric arc furnaces
- Reverberatory furnaces
- Crucible Furnaces
- Heat treatment furnace

The selection of furnace depends upon alloy quality and quantity of metal to be cast. Electric Arc Furnaces, Cupolas and Induction furnaces are used for ferrous metal. Reverberatory and crucible furnaces are used for producing aluminium, bronze and brass casting. Blast furnace is used to produce steel from iron ore.

The various processes involved in casting metals to the desired shape include:

- Process design
- Pattern making
- Directional solidification plan
- Preparation of sand mould
- Heating the metal in furnace to melt down to liquid stage
- Pouring the liquid metal into mould cavity
- Removing the cast from mould after solidification and cooling down
- Fettling the casting to remove components such as risers, runners, gate etc.
- Inspection of casting: Non – destructive and destructive process.

Process design:

The process design is done by design engineers considering

- Material composition of metal to be cast
- Volume of metal to be cast
- No of castings required
- Shrinkage of metal
- Directional solidification
- Position of chills to be provided
- Requirement of surface finish
- Machining allowance
- Internal shape in the casting and required surface finish
- Requirement of hardness on internal and external surfaces
- Position of Bush or basin, sprue or down runner, runners, ingates and risers.
- Design of pattern
- Design of core boxes
- Position of vent holes
- Moulding process (methods of moulding)
- Sand blasting and Fettling process
- Normalising the casting to suit machining requirements

The designer considers all above points and makes plan and drawings of entire process in 2d or 3D as he feels necessary. This design is performed by the "method department" in reputed foundry dealing with casting of various engineering components.

Pattern Making:

The process of making pattern is called pattern making. It is a special skill that is taught in reputed industries dealing with casting in foundry and technical institutes along with draftsman ship, tools and die making.

Pattern:

It is the replica of an actual object with additional systems like core prints and gating system such as runners, risers, gates, mounting boards and moulding box holding pins to enable moulding to take place. It is used to prepare sand cavity called mould to pour molten metal to get casting. The single piece pattern will look like as an actual object but differs due to draft and gating system. The patterns must have good surface finish. It is made with wood, metal, plastic material, araldite thermocol and wax. If made with wood its surface should be coated with varnish or paint to prevent warpage and have smooth finish.

COMPONENT

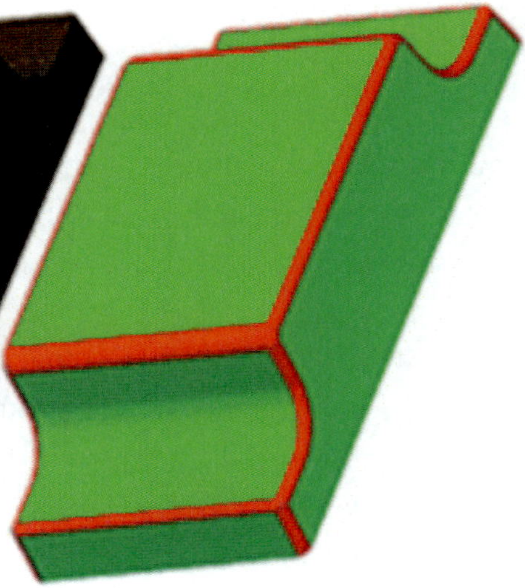

SOLID PATTERN

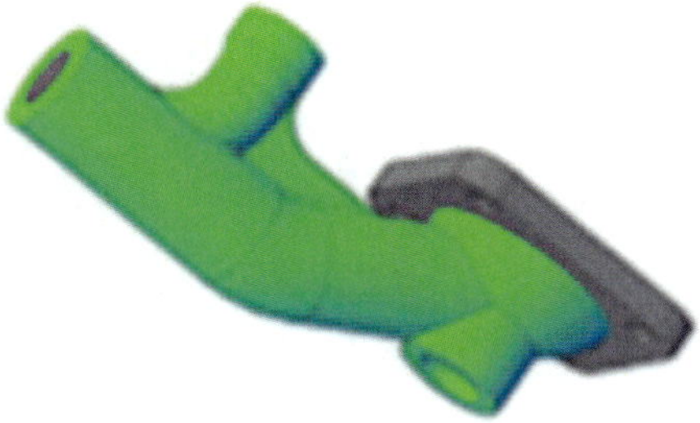

COMPONENT

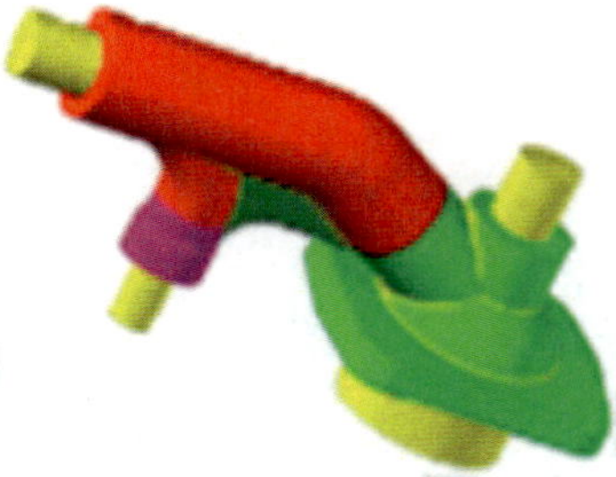

SPLIT PATTERN

The pictures of patterns above are shown without gating arrangement. You can compare the physical appearance of both components and patterns. The patterns with gating arrangement will appear more different than the component.

Warpage:

When wooden planks / patterns are stored for longer period, its moisture and sap dries out. The distortion / twist on surfaces during this period are called warpage. It is minimised by seasoning of timber / wood and applying paint on surfaces of patterns.

Gating system:

It is also called feeding system. It consists of pouring basin, sprue, sprue well, runner, runner well, gate and riser.

Mould:

It is a negative shape of the pattern, core prints and gating system made in moulding flask by ramming moulding sand all over the surfaces of pattern. It is made in two parts cope and drag.

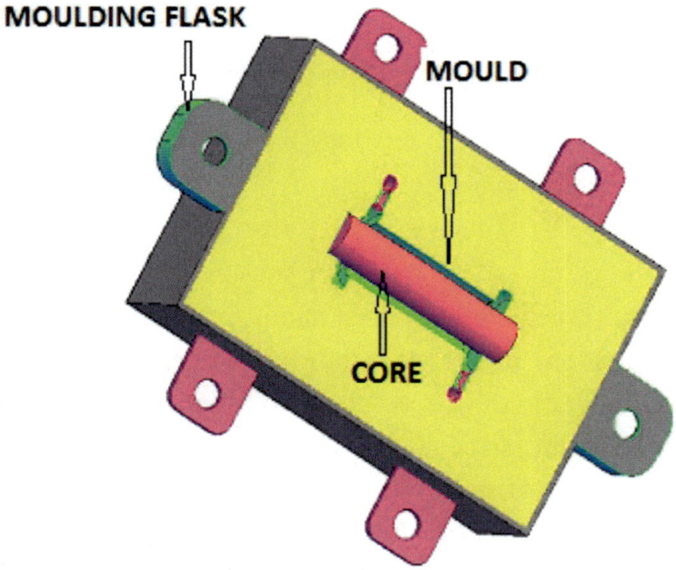

DRAG MOULD WITH CORE

Core print:

It is an extra projection added to pattern shape matching with internal profile. It forms a seat in the mould cavity when mould is formed by ramming sand mixture all over its surfaces in a moulding flask. This seat holds the core in cope and drag when set and locked together. The green coloured projected portion in pattern is core print

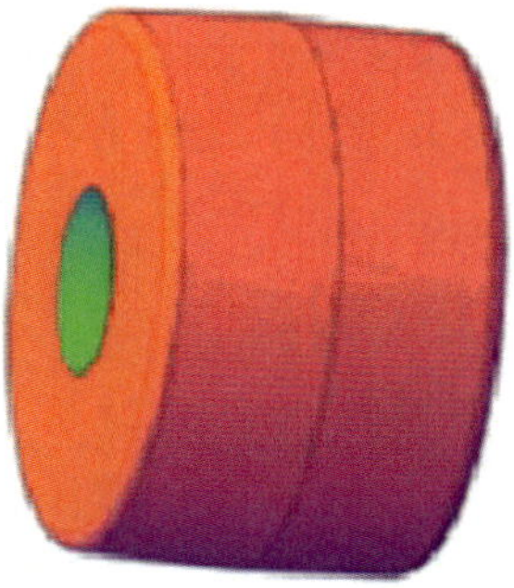

CASTING WITH HOLE IN CENTER

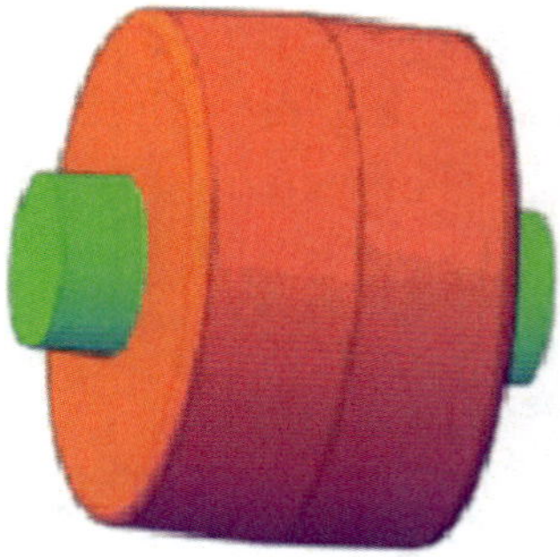

PATTERN WITH CORE PRINT

Core Box:

Core Box has two parts of body - external and internal. The external shape is given by pattern maker according to construction design with a prime consideration to accommodate internal profile of component and core print shape within it. The internal shape is a replica of internal profile of component with addition of shrinkage allowance and negative shape of core print attached with pattern. Its size and shape may not be the same as the internal profile of component due to some additions. The object taken out of core box after ramming sand mix is called core.

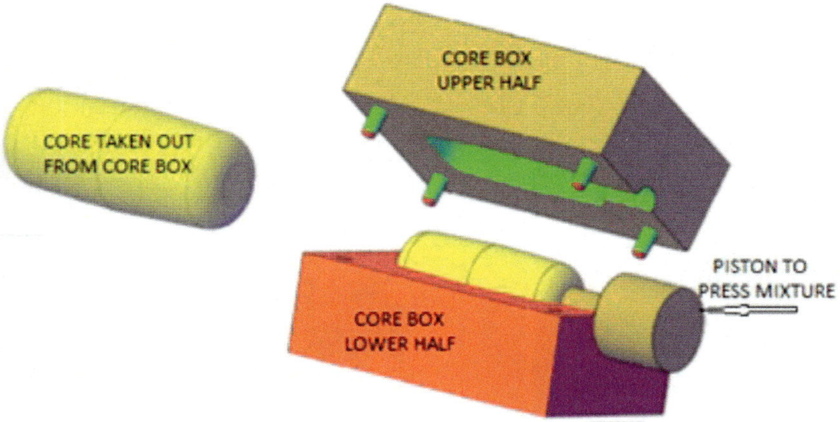

Types of core boxes:

- Internal profile core box
- External profile core box

Internal profile core box:

Core boxes have an internal casting shape. Core is nothing but a negative internal profile of casting. This enables mould to form the shape of the Core. The core boxes have a replica of internal profile of casting. The above drawing is the example of internal profile core box and core.

External profile core box:

There may be a profile in the component that lie above or below the parting line and is difficult to withdraw the pattern after ramming operation. In such case external profile may be made with core box to have casting for such profile. If there is any internal profile associated with such part of component, core box will have the replica of outer profile and core print. The proper draft and shrinkage allowance are to be provided as it is being done in case of pattern. When core is taken out from such core box it will resemble the part or whole cavity of mould with additional cavity to hold the core. The core box for internal profile of component will have negative shape of core print and replica of internal shape. Study the 3D drawings of component for making pattern and find the suitable parting line. See Pattern & mould making in foundry Vol – 2 for detail.

COMPONENT 'S MOSTLY EXTERNAL VIEW PROJECTED

COMPONENT'S INTERNAL CAVITY PROJECTED

Comparison of core box with mould

Sand mould	Core box
1. It is made of sand mixture. 2. It is formed by moulder after ramming sand mixture all-round the surface of pattern. 3. Its external shape is governed by moulding flask. 4. The cavity in the mould is the negative impression of pattern. 5. The product of mould is called casting. 6. The castings are obtained by pouring liquid metal in	1. It is made with wood, metal or plastic. 2. It is made by pattern maker with prime consideration to suit the internal shape of the component. 3. Its external shape is decided by pattern maker as per construction design, internal profile of component and core print. 4. The cavity formed in core box can be said as replica of internal profile of component and negative shape of core print. 5. The product of core box is called core.

cavities.	6. The cores are obtained by ramming core sand in cavity of core box.

Chapter-2

Explanation of Design Consideration

Pattern design considerations:

The important considerations in Pattern design are

- Shrinkage allowance: Solidification shrinkage and contraction
- Parting line
- Draft allowance
- Machining allowance
- Shake and distortion allowance
- Fillets and radius
- Gating system design: Positioning of Pouring spout, sprues, runners, gates and risers.

Solidification shrinkage and allowance:

When the molten metal is poured in mould cavity, it starts cooling from liquid stage to solid stage and then comes down to room temperature. The metal shrinks in two stages. It first shrinks while cooling down from red hot liquid to hot solid stage and then hot solid stage to room temperature. This first stage of shrinkage in casting is accomplished by providing risers.

The second stage of shrinkage, cooling down from hot solid to room temperature is compensated by addition of shrinkage allowance in patterns and core boxes. It shrinks from all directions during this period and hence allowance is provided in all directions to compensate.

The pattern is made bigger than the actual size of the component. The shrinkage allowance varies from metal to metal. The three stage of solidification has been explained by following drawings:

First stage of solidification:

It is decrease in volume of metal while passing from pouring temperature to freezing temperature.

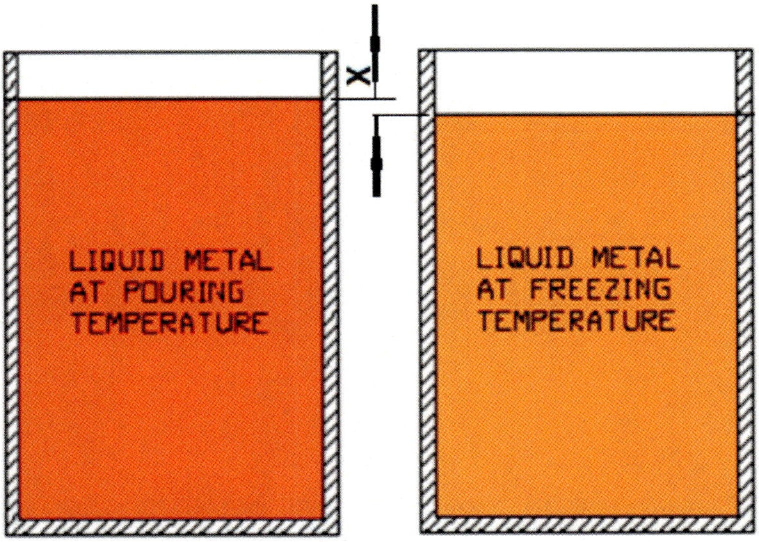

X= Shrinkage due to temperature freezing

Second stage of solidification:

It is decrease in metal volume due to solidification from liquid metal to solid metal

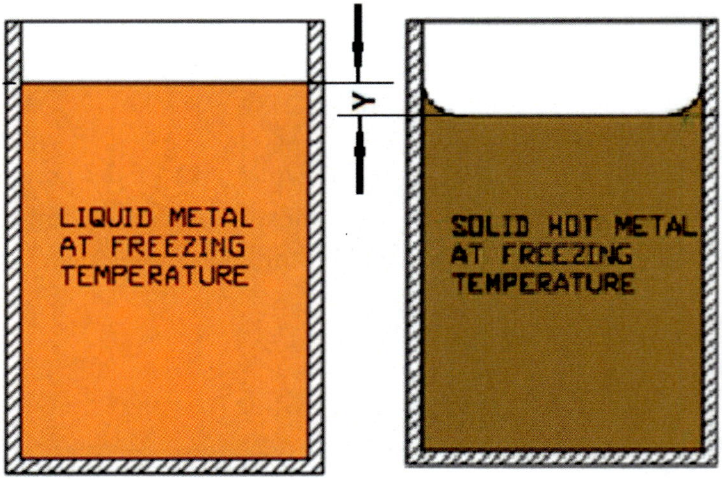

Y = Shrinkage during freezing from liquid to solid stage

Third stage of solidification:

It is decrease in volume of metal from hot solid stage to room temperature. It is taken care by method engineer and pattern maker.

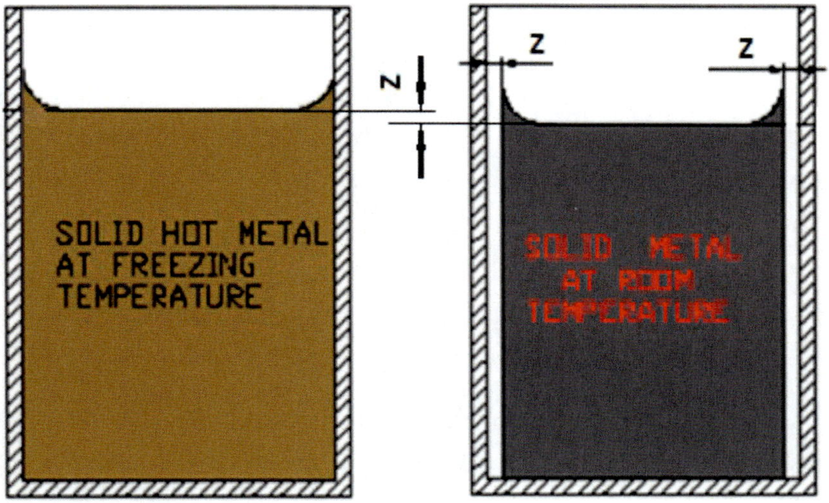

Z = Volumetric shrinkage at room temperature

Contraction allowances of metals:

It is also called pattern maker's shrinkage allowance as it is consideredand and added on each and every measurement of component drawing during pattern lay-out.

The allowances mentioned here are not appropriate. It depends upon compositions set by foundry. It is the responsibility of Research and development Laboratory to set the shrinkage allowance for method department.

Metals	Contraction %
Grey Cast Iron	0.7 to 1.05
White Cast Iron	2.1
Malleable Iron	1.5
Steel	2.0
Brass	1.4

Aluminium	1.8
Aluminium alloys	1.3 to 1.6
Bronze	1.05 to 2.1
Magnesium	1.8
Zink	2.5
Manganese Steel	2.6

Parting line:

It is very important reference line that divides patterns in two parts, cope and drag. It is also separation line/ plane of cope and drag moulds. It is taken as reference line for draft allowance. It may be in single plane or in different planes.

The selection of parting line plays a very important role in pattern design. It should be such that pattern could be taken out easily from mould cavity without breaking side walls, surfaces or edge of mould.

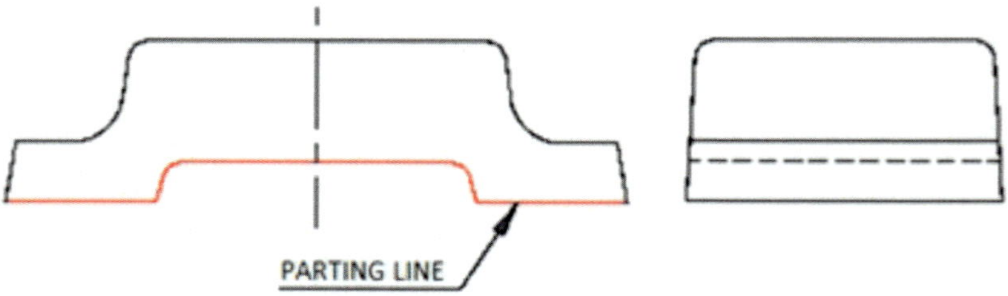

The above drawing is an example of single piece/ solid pattern. The parting line indicated by red line is not in a single plane. It is combination of vertical and horizontal plane.

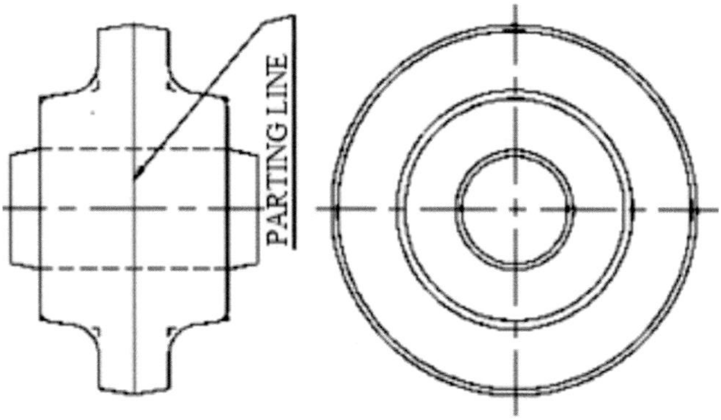

The red line in above drawing is the parting line that divides patterns in cope and drag. It is also a reference line for draft angle. The draft angle has been appropriately marked in drawing.

Draft Allowance:

The patterns are used for making sand moulds for pouring metals to get cast. The patterns and all accessories are to be taken out after ramming sand mixture all over its surfaces. The vertical surfaces and edges of mould should not break while taking out the patterns and accessories.

The necessary taper is provided on all vertical surfaces with respect to parting line /parting plane to facilitate easy removal from cavity without any damage on corners, surfaces and edges. The taper provided is 0.5 to 3 degrees depending upon requirement. The taper on vertical surface of mould with respect to parting line/parting plane is called draft allowance. The draft angle depends upon;

- Complexity of component
- Intricacy
- Moulding method
- Pattern's size and shape

The drafts have been shown by the drawing for better understanding.

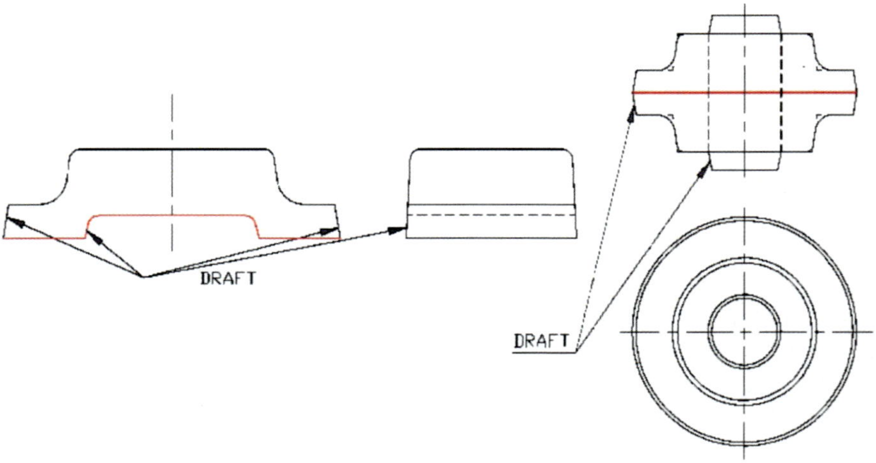

The red colour line is the parting line or parting plane in solid and split pattern. The draft lines with standard draft taper have also been marked.

Effect of draft, fillet and radius in mould cavity and casting:

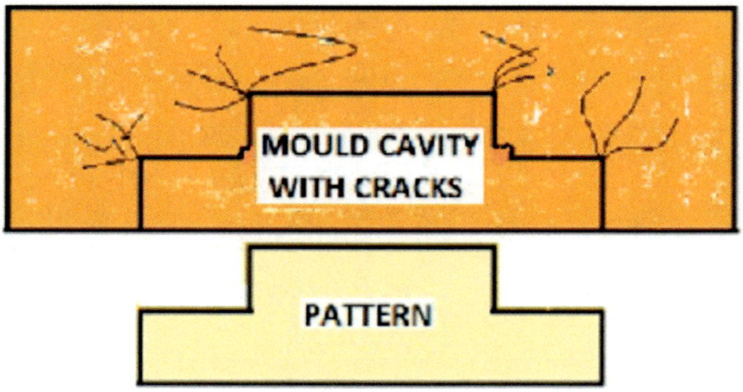

PATTERN WITH DRAFT, FILLET AND RADIUS

The pattern in above figure does not have proper draft angle, radius and fillets. The cracks have developed at the internal corners while withdrawing pattern from mould cavity after ramming operation. The moulds also has loose sand at external corners in the mould cavity. It will create metal penetration and sand inclusion in casting due to cracks and washing out of sand from loose corners.

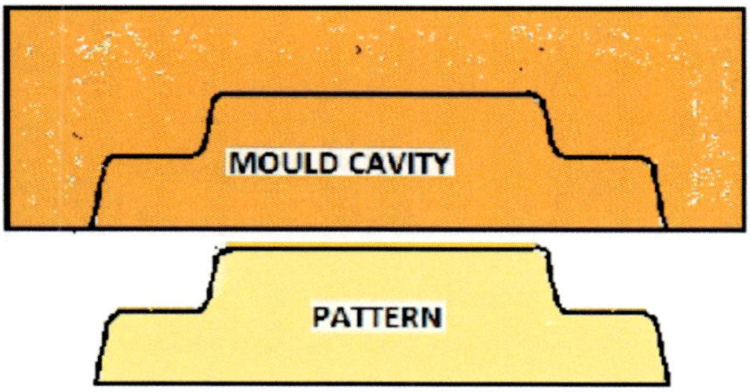

PATTERN & MOULD WITH PROPER DRAFT,FILLET AND RADIUS

There is no crack and loose sand in mould cavity, it is due to proper draft with respect to parting line, fillets and radius at corners. If the pattern is smoothly withdrawn from cavity, there is no chance of metal penetration or sand inclusion in casting.

Machining allowance:

The casting surface needs extra material over it to facilitate machining. The castings which are part of assembly will require better surface finish than non-assembly components. The Patterns are made bigger in outer size to have extra material on external surface where machining is required but inner sizes should be smaller than component where internal surface needs machining. Take the case of a simple cylindrical cast pipe where machining is required on external as well as internal surface. In this case, outer diameter of cast pipe should be bigger and internal diameter smaller than the machine finished pipe. Here core should be made smaller than the internal dia. of pipe and outer dia. of pipe should be bigger. Facilitation of this provision with Pattern and core box is called machining allowances. The extra material required on surface that is to be machined depends upon

- Method of moulding
- Size and shape of casting
- Degree of finish and tolerance required by customer.

The machining allowance has been indicated with hatched section in the above drawing.

Shake allowance:

Shake allowance refers to negative allowance. The pattern is wrapped all around the faces to facilitate easy removal from sand cavity after ramming. The cavity gets enlarged in this process. The pattern is made small in size to compensate this enlargement which is called shake allowance. There is no specific measure to compensate it. It depends upon individual skill of moulder and accordingly allowance is provided. It is required where machine moulding is not involved and manual work is done.

Distortion allowance:

During cooling of the mould, stresses developed in the solid metal that may induce distortions in the cast. It is also caused due to unequal cooling of different sections. This is more evident when the mould is thinner in width as compared to its length. This can be eliminated by initially distorting the pattern in the opposite direction. It also can be taken care by machining allowance.

REQUIRED CASTING

DESTORTED CASTING

DISTORTED PATTERN

Fillets and radius:

The fillet and radius play a very important role in patterns to prevent erosion of sand mix in mould cavity during poring of the molten metal. The fillet helps to slow down metal velocity making it more laminar. The rounding of internal corner of pattern (Intersection of inclined or vertical plane and horizontal) is termed as fillet whereas rounding of external corner (intersection of horizontal plane and vertical or inclined plane) is termed as radius. It has been explained by a sectional view of pattern in the next diagram.

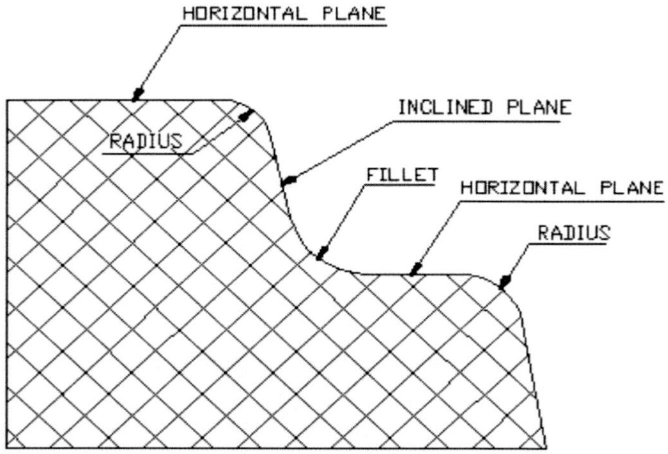
HORIZONTAL PLANE
INCLINED PLANE
RADIUS
FILLET
HORIZONTAL PLANE
RADIUS

Chapter-3

Nomenclature of Patterns According to Construction Materials

Category of Patterns for construction materials:

- Plaster of Perish Patterns
- Araldite Patterns
- Thermocol Patterns
- Plastic Patterns
- Wax Patterns
- Wooden Patterns
- Metal Patterns

Plaster of Perish pattern:

The body of pattern including pouring spout, runners, risers and sprue are made with plaster of perish. It is used for few accurate casings.

Araldite Pattern:

Epoxy resin and phenol formaldehyde as hardener is used for making the pattern with all accessories. The plaster of parish mould is first made by casting perish over wooden pattern. The releasing agent is applied over the entire surface of perish mould to get araldite shaped pattern released easily, then coating of epoxy resin (araldite) mixed with hardener is applied on surface of mould with ceramic wool / fibre as body reinforcing materials to get pattern shapes.

Thermocol Pattern:

The thermocol is used for making single large casting of unusual size. The sand mixer is rammed over it in the moulding flask. The thermocol burns out when molten metal is poured over it to get casting. The moulding process is same as other patterns.

Plastic Pattern:

In place of wood or metal, Plastic sheets are used for few patterns.

Wax Pattern:

The wax is used to make patterns. The ornaments are cast with the help of wax patterns. The casting process is also called lost wax process.

Wooden Patterns:

Wooden Patterns are generally used for large patterns as well as for medium size repetitive castings. The seasoned teak wood is used for this purpose as deformation in dimension and shape are very less. It is used to cast shapes where dimensional accuracy is required. The chip variety of wood is also used for single piece large casting where accuracy is not very important.

Metal Pattern:

The metal patterns are used for repetitive castings for prolong period and machine moulding process. The aluminium alloys, cast iron and steel are used for making it.

The patterns are first cast with the help of wooden patterns called master pattern made with necessary machining allowances. The casting is machined with required grade of finish, shrinkage allowance, draft and machining allowance that are required by patterns. All accessories such as pouring spout, runners, risers etc. are also cast and machined before mounting on cope and drag board.

Machines required for making metal patterns:

- CNC Milling machine
- CNC Grinder
- Copying lathe
- Metal cutting Bandsaw
- Welding machine
- Drill machines

Hand tools and accessories

- Drills bits of various sizes
- Taps of various sizes
- Clamps
- Hexagonal and Allen bolts of various sizes.
- Vernier calliper
- Inside and outside callipers
- Digital callipers
- Divider
- Scriber / marker

Chapter-4

Category of Wooden Patterns

Category of wooden Patterns considering design, construction and use:

- Solid Patterns
- Split Patterns
- Cope and drag pattern
- Mounted Patterns
- Gated pattern
- Self-core pattern
- Follow board Pattern
- Match plate pattern
- Loose piece pattern
- Skeleton Patterns
- Sweep Pattern

Solid Pattern:

The solid pattern is the single piece pattern having parting line at one convenient end face or plane. It is used for making large casting. It is also called loose pattern or single piece pattern.

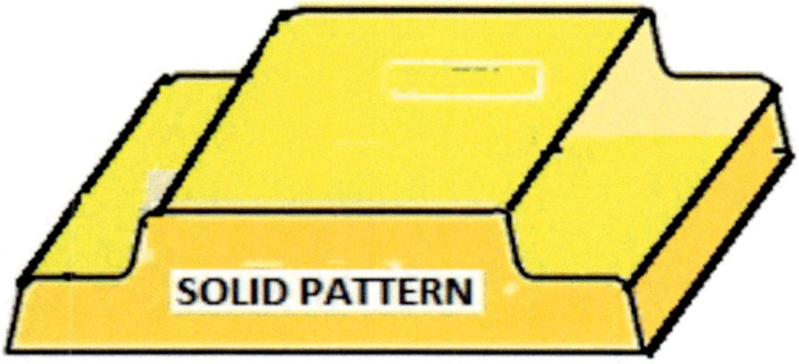

Split pattern:

The split patterns are made in two pieces split at the parting line. They are joined together with dowel pins to eliminate mismatch during moulding. The parting plane may be a flat or irregular surface. It may be split in more than two parts if shape is more complex to facilitate withdrawal. It is used for few medium size castings.

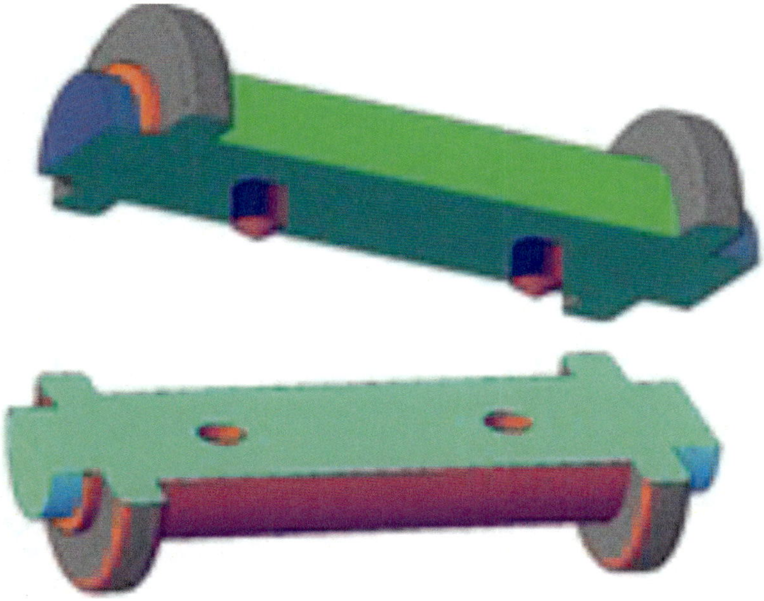

Cope and Drag Pattern:

In case of large casting, patterns are split in cope and drag. The cope and drag parts are mounted on two separate mounting boards with proper gating arrangements and provision for centring moulding flask to eliminate possibility of mismatch in cope and drag.

Mounted pattern:

When split patterns are mounted on wooden boards with provision to hold moulding flask in an appropriate position to eliminate mismatch, it is called mounted pattern. The mounted patterns are used for mass production having more stringent and accurate castings. It is also called cope and drag pattern. The gating system also may be attached with cope and drag mounting board with patterns.

Gated pattern:

When more than two small and medium size split patterns are mounted on cope and drag mounting boards with gating arrangement, it is called gated pattern. The gated patterns are used for mass production of small casings for regular use. These are used for making moulds by automatic moulding machines.

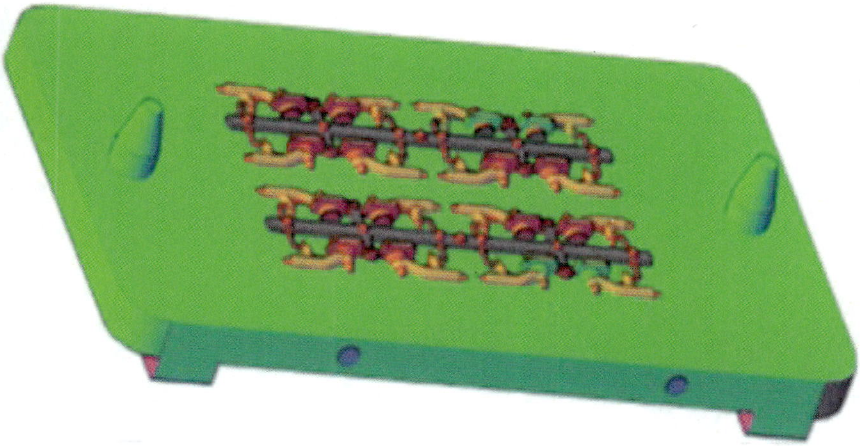

Drag Patterns Mounted on Mounting Board with Gating System:

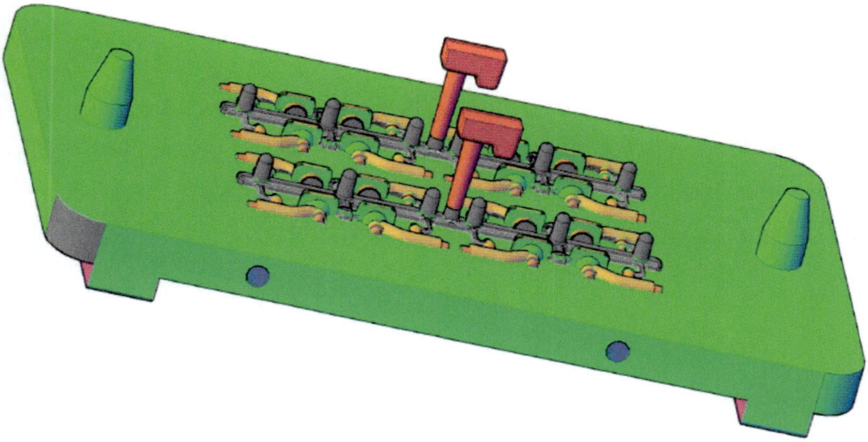

Cope Patterns Mounted on Mounting Boards with Gating System and Risers

Self-core pattern:

In this case the profile of pattern having cavity lies on parting line and serves as core box. This cavity will form core (projected sand body) during moulding process. This projected body will sit and occupy space in cope mould like a core to form cavity in the casing.

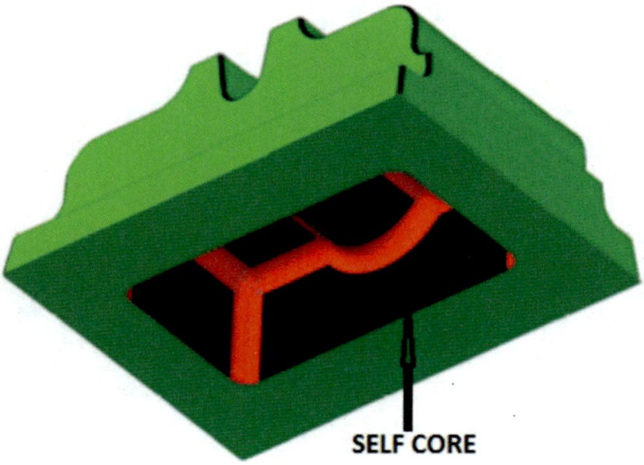

Follow board pattern:

It is a device that will follow the pattern to facilitate moulding. It is required for pattern of thin cross section or having irregular parting surface that can't be moulded with other pattern design. See the pattern on follow board given below and visit the pattern making and moulding exercise to clear concept.

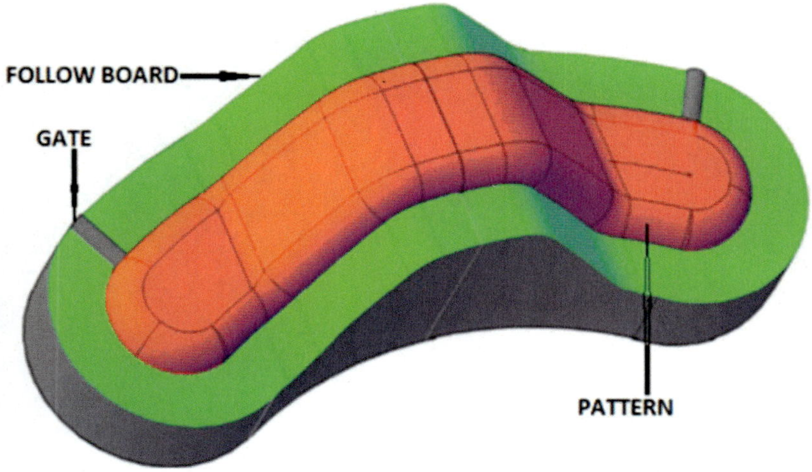

Match plate pattern:

The cope and drag patterns mounted on opposite sides of a mounting board is called match plate. The gates, runners, risers are also mounted on match plates. The moulding is automated by machines. It is used for mass production of small castings.

Loose piece pattern:

When some part of pattern lies above or under the parting line, making it difficult to withdraw from mould cavity, that part of pattern body is made loose. The pattern is withdrawn from cavity after moulding leaving the loose piece to remain inside cavity. It is removed by moving it in a safe direction to withdraw. This type of pattern with loose component is called loose piece pattern. The green and red coloured piece will be withdrawn from cavity after taking out the pattern.

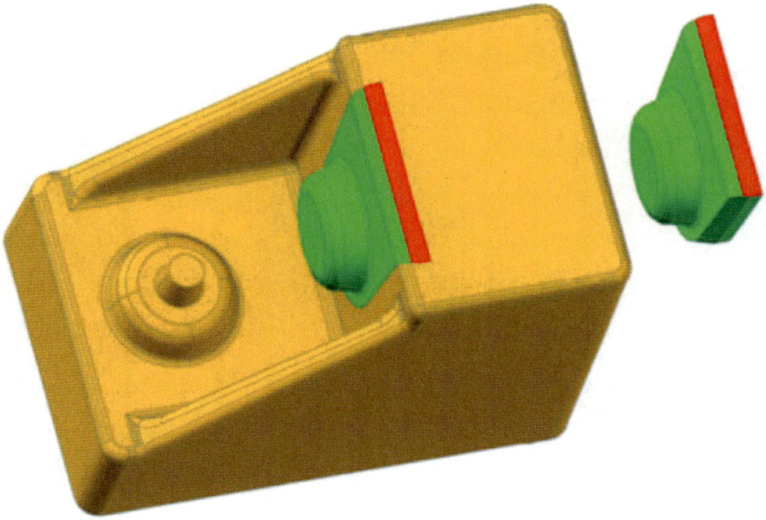

Skeleton pattern:

The skeleton name is given to pattern on the basis of construction. The structure of skeleton pattern consists of ribs and bones frame like as our body. The surface is made smooth all over the structure by a layer of wood.

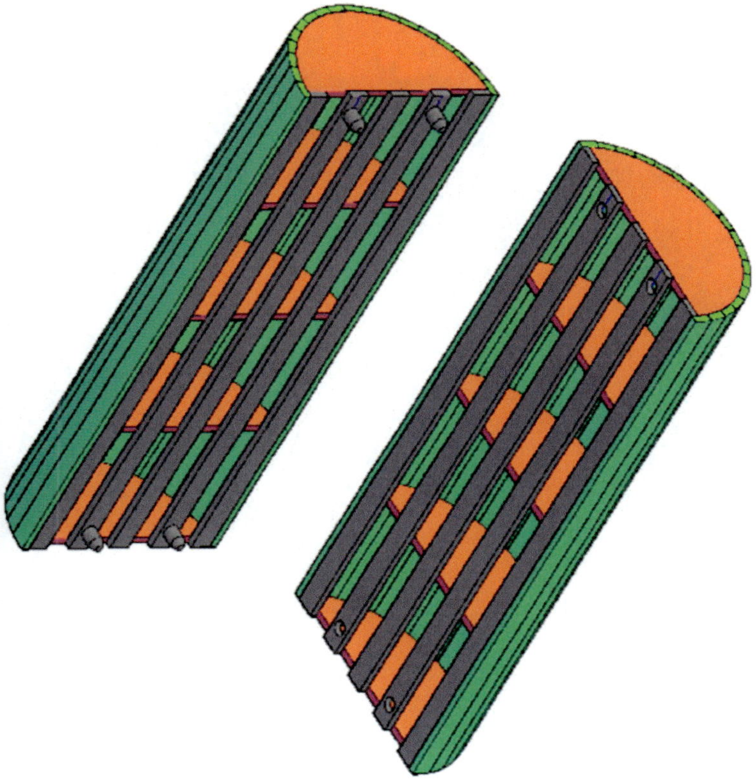

Sweep pattern:

The word sweep means cleaning. We clean our house with broom similarly the mould cavity is made by sweeping the rammed sand mixture with the help of template. The process is followed for large single casting to save the cost of making expensive pattern. It is generally used for castings of circular nature.

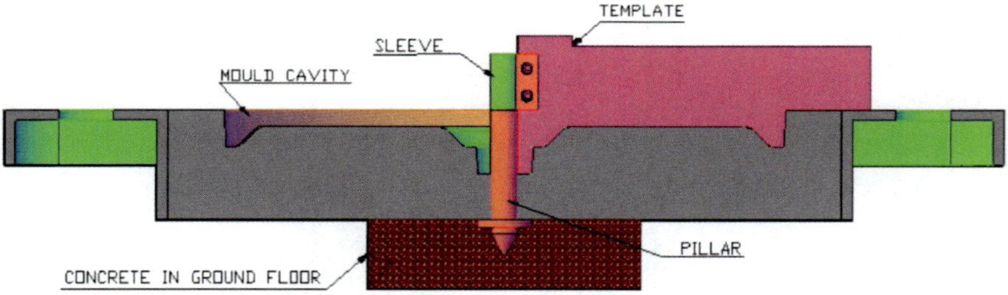
TEMPLATE
SLEEVE
MOULD CAVITY
PILLAR
CONCRETE IN GROUND FLOOR

Chapter-5

Pattern Construction Decisions

&

Steps of Pattern Making

Decision on Pattern construction:

The decision on construction of pattern is taken by pattern makers / method engineer considering following factors

- No of castings required
- Size of casting
- Shape of casting (intricacy of shape)
- Accuracy and finish required
- Material composition
- Type of moulding and casting method

Suppose a very large size single piece casting is required. The pattern maker may decide to have an economical construction. If it can be a single piece pattern, he will prefer skeleton pattern or sweep pattern. He will also consider accuracy and feasibility of moulding such shape.

In case of mass production, mounted pattern on mounting board will be made. In certain cases, where single piece is not possible split pattern will be made for few castings. The categories of patterns are also decided considering above factors.

Steps of pattern making:

- Thorough study of drawing to have clear concept of component to be cast.
- Consider machining allowances on surfaces according to requirement on individual surface for perfect application.
- Contraction allowance to be considered according to cast metal
- Parting line to be decided for proper withdrawal of pattern from mould cavity.
- Draft allowance and shake allowance
- Position and size of sprue, sprue base, pouring basin, runner, risers etc.
- Preparation of lay-out after calculating pattern sizes.

Layout:

It is a drawing of component made with 1:1 scale on lay-out board made of plywood board or thick card board with necessary allowances and additions such as contraction allowance (to compensate shrinkage in metal from red hot solid to room temperature), machining allowance, draft and core prints. The parting line which is very important aspect of pattern making is decided and indicated clearly on drawings where ever it can be. The pattern makers use contraction scale for placing measurement on layout board but if it is not available they can calculate the required sizes mathematically considering percentage of shrinkage.

The construction method detail will be an added advantage if plotted clearly. The fillets and radius should also be marked on layout drawings.

The runners, ingates, risers, sprue, poring basin, sprue well and chock also should be plotted clearly.

Contraction scale:

The contraction scale is also called pattern maker's scale or shrinkage allowance ruler. It's ruled marking are made with larger spacing according to shrinkage taking place in different metals. The rules side is marked with metal and its shrinkage %. All metals contract by an amount distinct from others. The contraction scales are available for the materials to cope up with the foundry men's requirement.

Chapter-6

Elements of Gating System, Types of Gating System, Feeders & Design Considerations

Elements of gating system:

Poring basin:

It can be defined as receiving end of molten metal from ladle. The molten metal enters into sprue from pouring basin. It should be capable to hold molten metal and supply uninterruptedly to sprue during pouring.

The dross and slag that will float on surface of molten metal is maintained in it. It only allows clean metal to enter into sprue. One wall of pouring basin should be made 45 ^{0} inclined to horizontal plane. The molten metal is poured on this wall to reduce momentum and vortex formation.

The pouring basin depth should be 2.5 times the sprue entrance diameter. It should be deep enough to hold molten metal and maintain continuous stream of metal in sprue. The entrance of sprue from basin should be smooth having minimum radius of 25 mm.

Sprue:

It is circular section entry path for the molten metal from pouring basin. It has larger cross section at top and smaller at the bottom. Its area is quantified by choke area and gating ratio. The size and shape of it should be able to minimize dross (Mineral waste that forms on the surface of molten metal as a result of oxidation) formation and achieve the required rate of metal flow with no turbulence. It must supply sufficient metal to mould quickly before solidification time.

The cross section area of the sprue exit or bottom is calculated from the choke area and gating ratio. The area of the sprue top can be calculated using mass and energy balance equations to prevent flow separation in the sprue.

Sprue Well:

It should ideally be shaped **cylindrical.** The diameter should be twice that of sprue exit and depth twice that of runner. It should be located beneath sprue and under the runner. It is designed to restrict free fall of metal in runners and direct it in right angle direction toward runner. The design reduces turbulence and air aspiration.

Runner:

The runner is the horizontal distribution channel that receives the molten metal from sprue and delivers it to gates and finally to casing. One runner can be used for simple casting but more than one is to be used for complicated castings. It is also designed to trap dross, impurities and prevent it from entering the gates and finally to mould cavity.

The dross traps also called runner well is designed at the end of runners for this purpose. The runner should project above the gates to ensure flow of pure metal to the cavity as dross and impurities float on surface of metal.

The cross section area should be greater than sprue exit. Its function is to slow down the speed of molten metal after free fall from sprue. Its design should allow filling it completely before ready to deliver metal to ingates. Lowering down of cross section of runner is important to ensure smooth flow of metals to ingates where more than one ingates are provided.

Runner well:

It is designed at the end of runners to trap the dross and impurities and prevent it from entry to mould cavity.

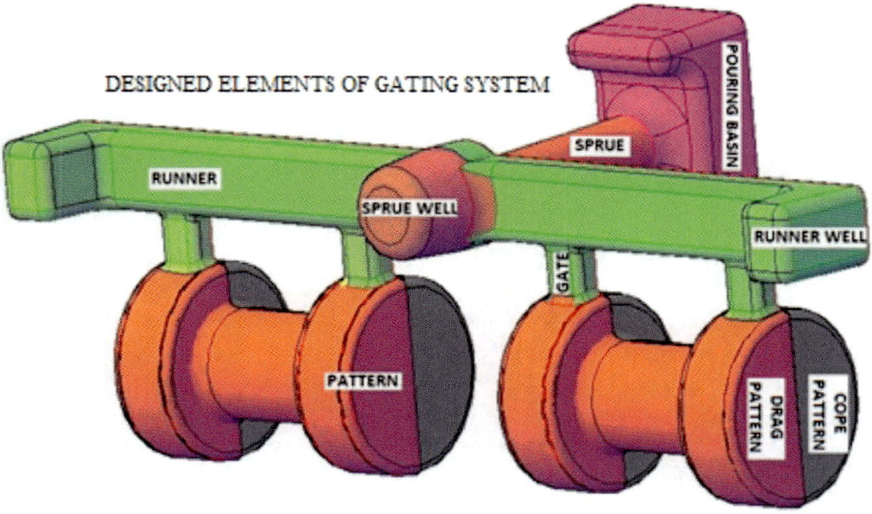

Fillet at the junction of sprue end and runner:

The end of down sprue that will touch the parting plane and meet with runner surface should be rounded up to prevent sand erosion from corner and inclusion in metal. The loose piece design of end point of down sprue has been shown in the above drawing to serve this purpose.

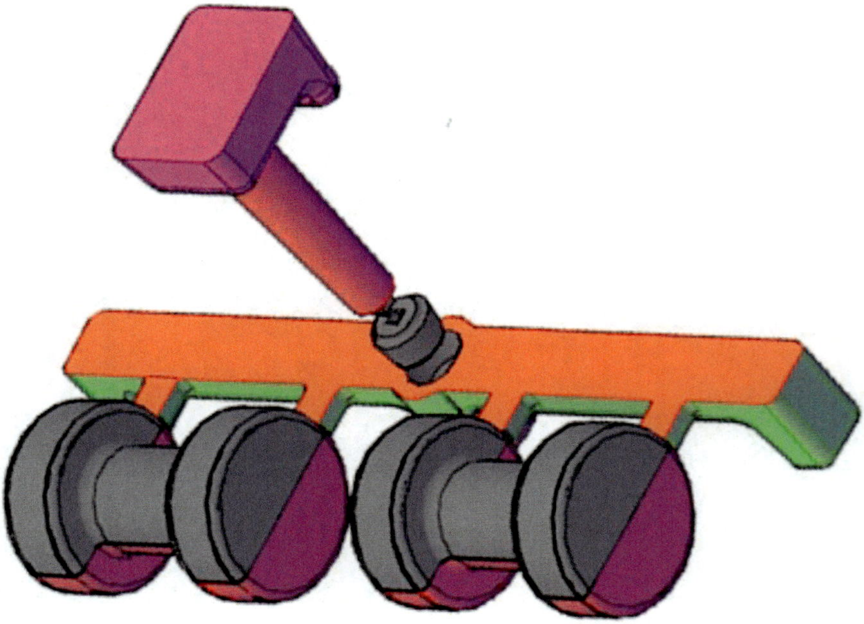

Gate /Ingate:

It is the connection between runner and mould cavity to deliver molten metal to create casting. It should be located towards the heavy section of casting. It should be at higher level than runner depth to insure impurity free delivery.

Ingate choke:

It is circular blind ended shape designed at the exit end of sprue to entrap dross and impurities prior to entry in mould cavity.

Types of gating system

The gating system is classified as follows depending on their connection with pattern and finally location in the mould cavity

- Horizontal

- Vertical
- Top
- Bottom
- Side gating/ middle gate

Horizontal gate:

It is suitable for flat ferrous casting filled under gravity. It is used for gravity die casting of nonferrous metals.

Vertical gate:

It is suitable for tall castings, shell moulds and die castings.

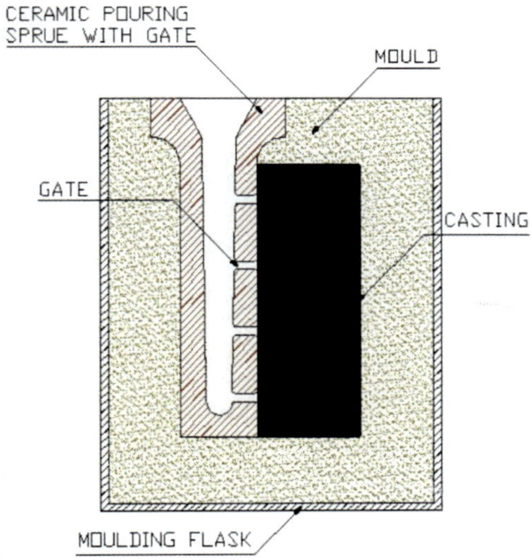

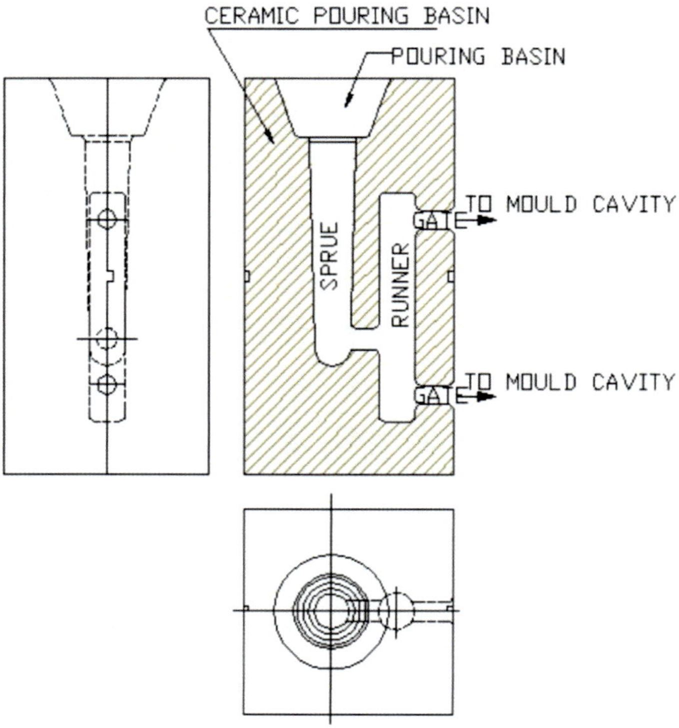
CERAMIC POURING BASIN
POURING BASIN
SPRUE
RUNNER
GATE
TO MOULD CAVITY
GATE
TO MOULD CAVITY

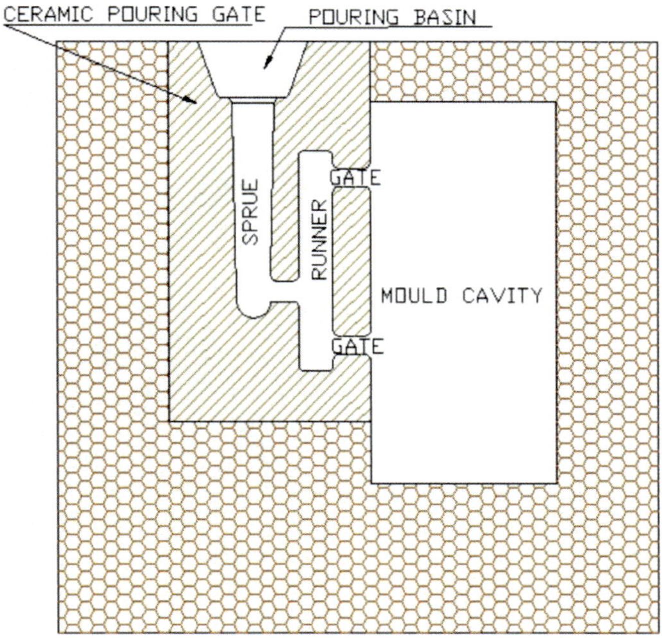

Top gating:

In this system molten metal enters at top of the mould cavity. It promotes directional solidification from bottom to top of casting.

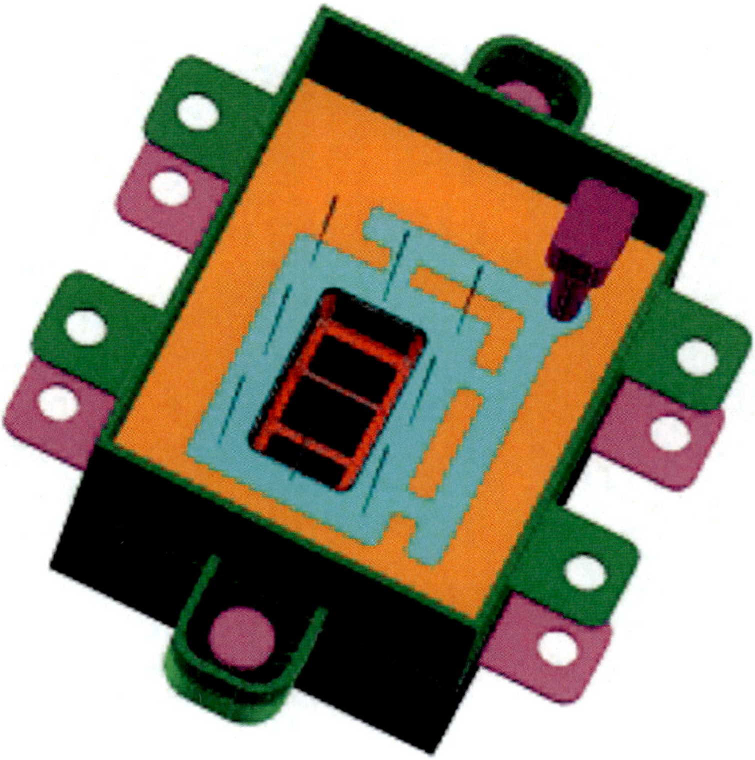

The cope mould flask with pouring basin, sprue and vent rods is clamped over drag for ramming sand mixture. The drag mould still holds single piece self-core pattern, runners, ingates and sprue well.

After moulding the cope mould will be taken up vertically from drag. It will have pouring basin, sprue, vents holes, and self-core. Now the pattern with gating arrangement will be withdrawn from cavity. The drag part will have mould cavity with top gates

Bottom gating:

In this system the molten metal enters from bottom of mould cavity and fills it gradually and slowly from bottom to top. The bottom pouring refractories bricks also can be used.

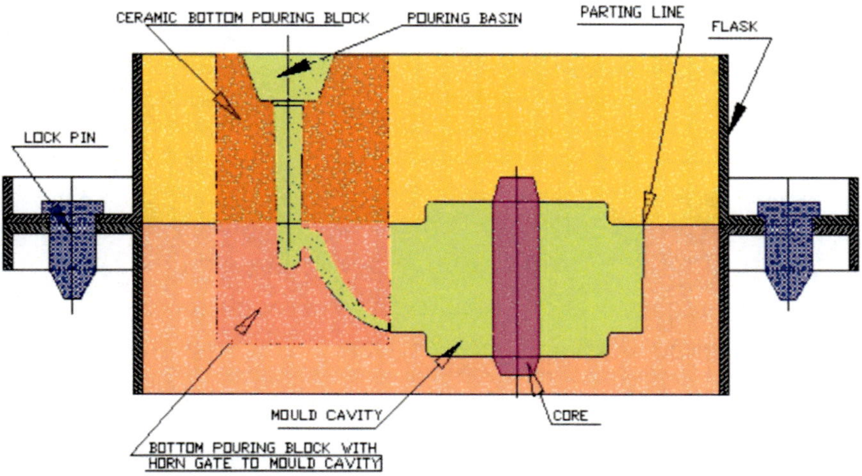

Side gate/ middle gate:

It is the combination of top and bottom features

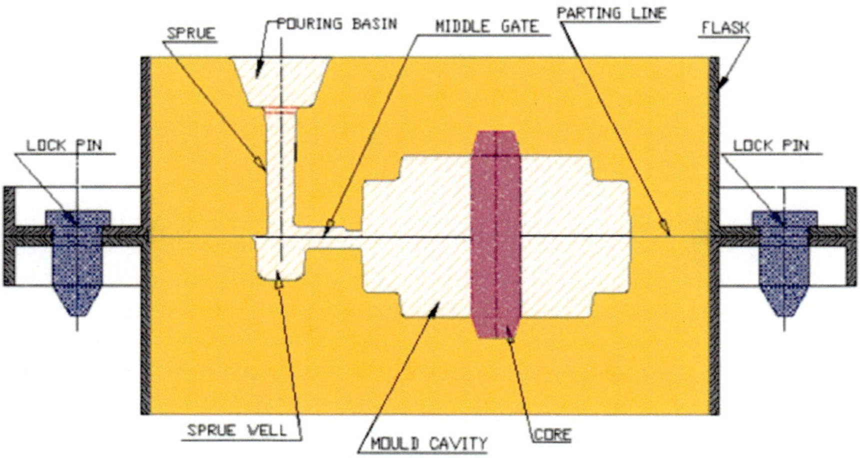

Gating system design:

The gating system is one of the most important aspects of pattern design. It is the path through which molten metal inters and flows into mould cavity.

There should be always liquid metal in the gating system to accomplish solidification shrinkage and prevent porosity formation in casting. You must calculate the volume of all areas of casting accounting for shrinkage.

The gates are usually attached to the thickest part of the casting to assist in controlling shrinkage. In large casting multiple gates or runners are required to fill mould cavity through more than one point. The proper design of gating system is very important to

- Prevent casting misrun due to premature solidification. It can be done by designing short flow path. The short flow paths and fast metal flow should be considered without any obstruction and direction change. It will reduce turbulence.
- Control the turbulence and trapping dross.
- Control the speed of liquid.

We should also consider the following to have defect free castings.

- Gating system should be relatively heavy to make solidification of metal slow. In this case there will be fast accomplishment of solidification shrinkage.
- Fillets should be provided where it meets the casting / mould cavity/ pattern.

- The existing end of sprue must be rounded to eliminate erosion of sand.
- All vertical straight and curved surfaces must have proper draft with respect to parting plane to facilitate easy removal of components form mould cavity.
- Gate should be sufficiently away from sprue.
- It should not be placed opposite to core.
- Runner, runner well, ingate, ingate choke, sprue, sprue well, risers and vent holes are very important compartment of gating system.
- The gates with longer length and wide passage are bad design whereas short length with normal cross section is good design.

Turbulence:

It is defined as fluctuating flow of molten metal with disturbances while poured in the mould cavity through gating system. There are two types of turbulences

- Bulk turbulence: It is ratio of inertia to viscous pressure in fluid
- Surface turbulence: It is ratio of inertia to surface tension pressure in a fluid

Bad effect of turbulence:

- Mould erosion (wearing out on internal surfaces of mould cavity)
- Creation of porosity in the casting

Vent holes:

These are small tiny holes made in cope parts of mould. It starts from cavity surface/parting surface of cope and goes up to top surface of mould. It is designed to provide passage to escape gases that form during pouring liquid metal in mould cavity.

Risers:

The risers are reservoir of metal that feeds the casting in mould cavity during solidification shrinkage from red hot liquid to hot solid stage. The shrinkage must be calculated to ensure the volume of material needed to accomplish solidification shrinkage in casting. The design of riser should accomplish following features.

- It should not solidify before the casting. The shape of riser should ideally be spherical to have lowest surface area to volume ratio, but it is not possible to connect with casting cavity and hence cylindrical shape is preferred.
- The volume of riser should be large enough to compensate the volumetric shrinkage in cavity during solidification.
- There should not be any possibility of hot spot to develop shrinkage porosity.
- The riser should be connected to feed thickest and largest section of casting where it is most wanted.
- The pressure head from the riser should supress cavity formation.

Solidification shrinkage of metals:

Metals	Solidification Shrinkage %
Grey cast iron	2.5 to 1.6
White cast iron	4.0 to 5.5
Ductile cast iron	4.5 to 2.7
High Carbon Steel	4.0
Copper	4.9
Aluminium	6.6
Aluminium alloys	-
Bronze	-
Magnesium	4.0 to 4.2
Zink	3.7 to 6.5
Manganese steel	2.6

Type of risers:

The categories of risers are classified on the basis of their position and exposure to atmosphere.

When it is located on the casting, its called **top riser**, but when it is located by the side of casting in mould cavity on parting line, it is called **side riser.** The categories of risers are called **open risers** when their top surfaces are exposed to atmosphere.

The top risers take less space in flask and are advantageous due to shorter feeding distance. The open risers are bigger in length as it has to go up to top surface of cope mould either from top of mould or from parting surface. It loses more heat due to exposure in atmosphere from the top surface. The top risers are almost always dead risers as it receives material through mould. In this case the molten metal first fills the mould cavity then metal passes to riser, there is no other source to feed it during solidification.

When it is being fed through gating system, its termed as live riser. In this case material in gating system will also feed it during solidification and shrinkage in casting will be accomplished for longer periods of time.

When riser completely remains in side cope or drag part of mould it is called **Blind riser.** The blind risers are almost always live risers. The vent holes are provided wherever it can be to allow for gases to escape that form during pouring.

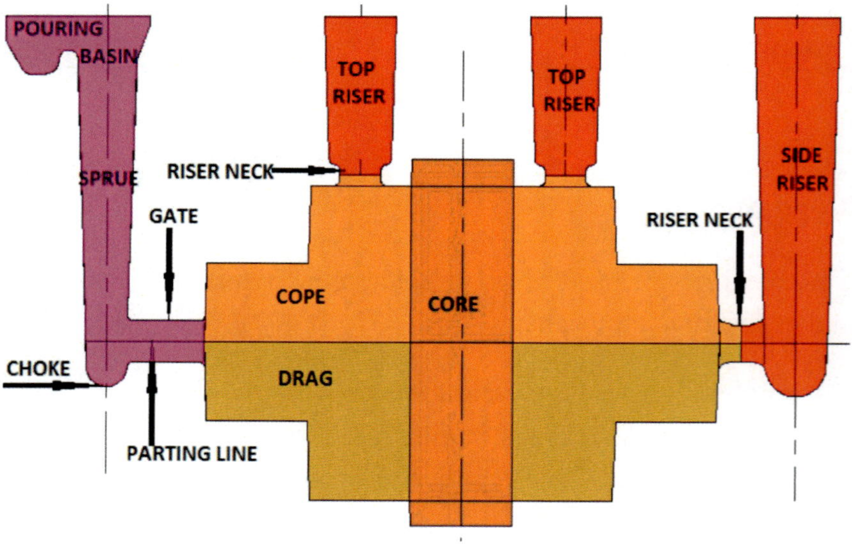

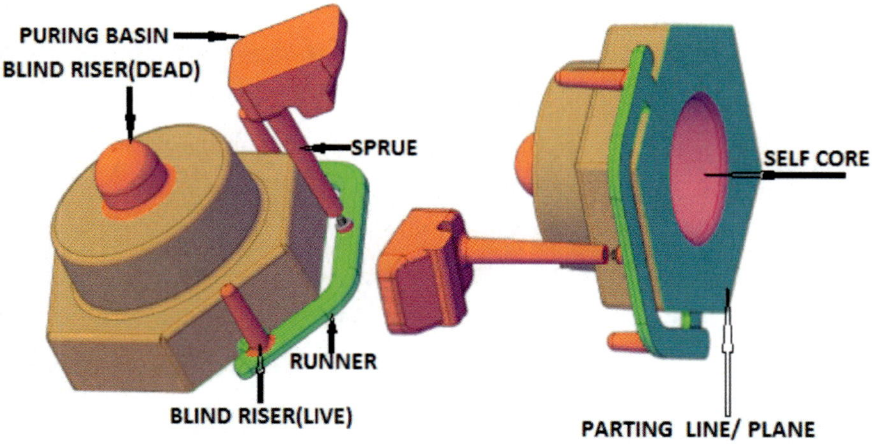

Advantage of Top riser:

- It has advantage of feeding casting by pressure (by weight of the metal).
- It occupies less space in flask.
- Feeding distance is shorter.

Advantages of blind risers:

- The blind risers are hemispherical at top and as such have higher volume to surface area ratio.
- It is not exposed to atmosphere and is attached to gating system; hottest metal remains in the riser to promote directional solidification.
- It is smaller than open and side risers that are going up to the top surface of cope mould.
- It can be located in any position in mould cavity.
- Removal from casting is easy.
- The cooling rate is slow and hence it can accomplish solidification shrinkage for longer period.

Design considerations:

- Expected shrinkage in casting must be calculated.
- Riser should be at higher level than the section of mould to be fed.
- Wherever possible bottom should be given hemispherical shape to keep it in liquid state for longer period.
- The neck of riser that connects to metal should be properly designed considering V/A ratio and allow the required liquid metal to flow in the cavity as needed.
- The square and rectangular neck length should be approximately 30% of riser diameter.

Locate and design riser to connect and feed heavy section of casting. Never feed heavy section through lighter section.

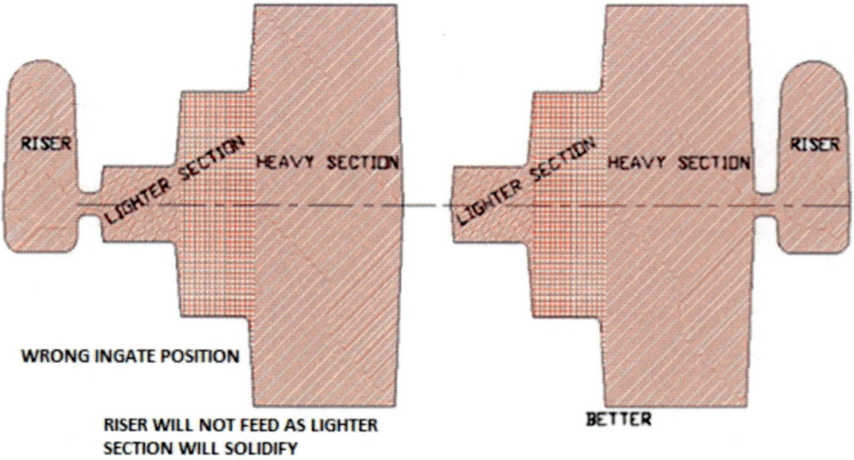

Make a provision to insulate risers by insulating or exothermic riser sleeve. The top surface of open riser is also insulated with insulating or exothermic compound. Along with shape, size and location grouping of castings should also be decided. The connection between riser and casting should not solidify before casting.

Connection between riser and casting:

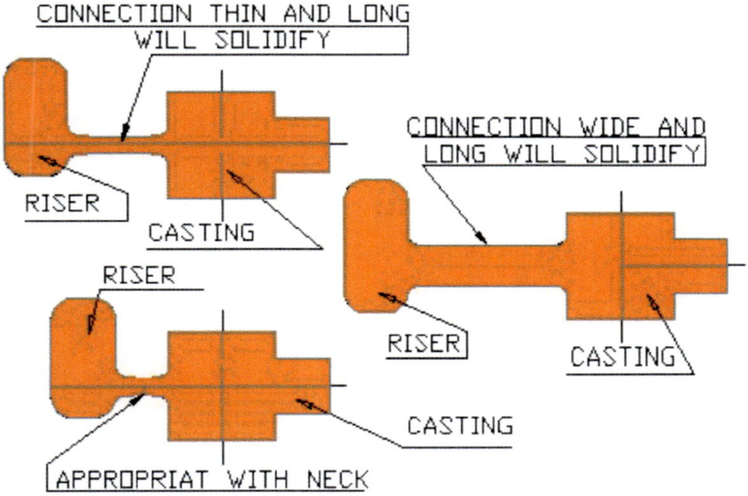

Preventing premature solidification of metal inside open riser:

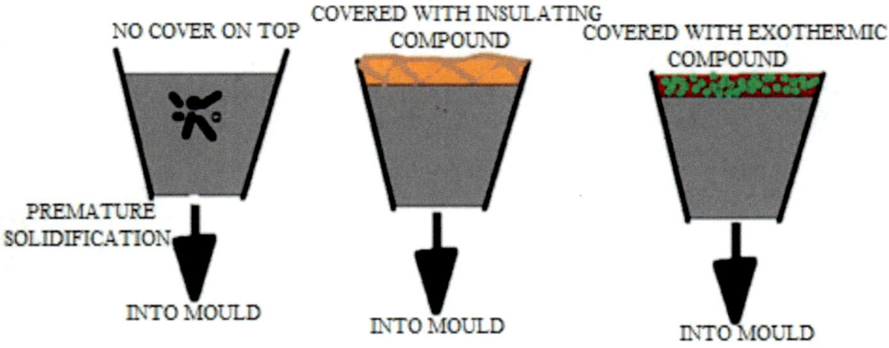

The heat loss from risers is also reduced by use of insulating or exothermic rise sleeves.

V/A Ratio:

It is the ratio of volume to surface area. The consideration of V/A is very important while designing gate and riser. When molten metal starts loosing heat, gradual reduction in temperature takes place.

The surfaces of mould cavities are at room temperature which is at much lower temperature than superheat. The molten metal when comes in contact with surfaces of mould cavity after pouring, starts loosing heat and a thin skin of solid metal is first formed on the surface between mould wall and casting. The thickness of skin increases progressively toward the centre of liquid mass.

The section in casting with low volume to surface area will solidify faster than section with higher volume to surface area. The proper consideration of V/A will eliminate premature solidification in casting.

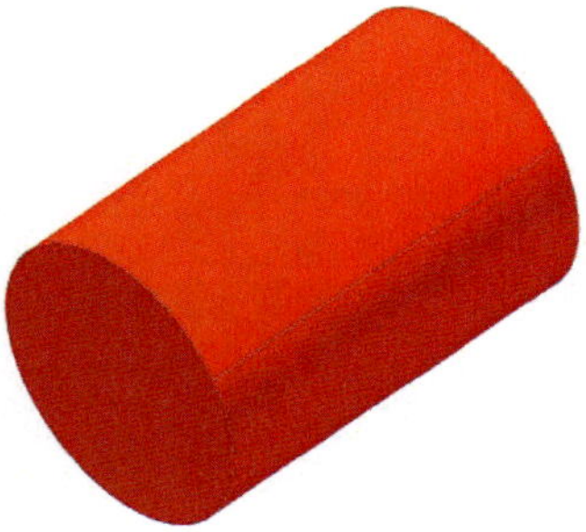

Let us consider above cylinder has 10 cm radius and 30 cm length. Its volume will be = 3.142*10*10 * 30 = 9426 cubic cm

Surface area will be = 2*3.142*R(R+H) =2*3.142*10*(10+30)

= 2513.6 square cm (R= Radius, H = Height)

V/A Ratio of cylindrical shape =9426/2513.6 = 3.75

L et us consider the above rectangular block with length =35 cm and thickness = 10 cm, width =W cm and volume = 9426 cubic cm. The value of W will be = 9426/35*10 = 26.93143 cm. The surface area of this rectangular block will be = (L*W*2) + (L*T*2) + (W*T*2) square cm

= (35*26.93143*2) + (35*10*2) + (26.93143*10*2) square cm = 3123.829 square cm

V/A Ratio of rectangular shape = 9426/3123.829 = 3.01745

The rectangular block having low V/A than Cylindrical block will solidify faster than cylinder.

Solidification:

The probable way of solidification of casting in mould cavity is

- Progressive solidification
- Directional solidification

Progressive solidification

When molten metal is poured in the mould cavity, solidification starts from the surface of mould. A thin layer of solid metal is formed in between mould walls and liquid metal. It then progresses gradually toward centre of casting metal mass. It is a natural self-solidification called progressive solidification.

Directional solidification:

When liquid metal is poured in the mould cavity, it starts losing its temperature. The moment it comes in contact with mould surface transformation of heat starts. The solidification starts from the farthest end of the casting and advances gradually towards the sprue / risers as we planned.

Example of progressive and directional solidification:

The transfer of heat energy from liquid hot metal to atmosphere is prime criteria behind solidification process. The characteristic of liquid is to flow from higher level to low level, similarly heat energy also flows from higher temperature to low temperature.

The pouring temperature of liquid metal is much higher than mould surface which is at room temperature. The moment liquid hot metal comes in contact to mould surface, the transfer of heat energy begins. The heat is continuously being transferred to atmosphere through sand grains.

A thin skin of solid metal's layer is formed in between hot liquid metal and all round mould's contact surfaces. The heat energy is transferred from liquid metal to solid skin, solid skin to sand grain and finally to atmosphere through moulding flask.

In this process, the solidification begins fast at farthest end from heat energy source and in thin section of casting. When it progresses perpendicularly towards centre of cast hot metal mass, it is called progressive solidification. When solidification path is diverted towards risers and sprue by proper design it is called directional solidification.

The different level of solidification has been explained here with the drawings

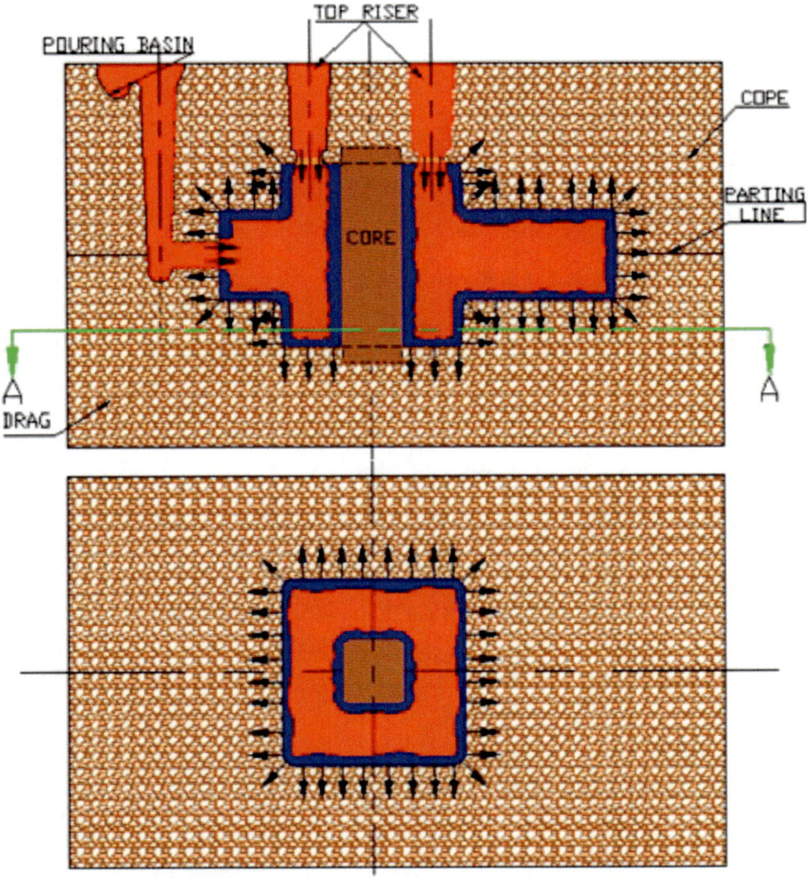

PROGRESSIVE SOLIDIFICATION AT LEVEL A - A

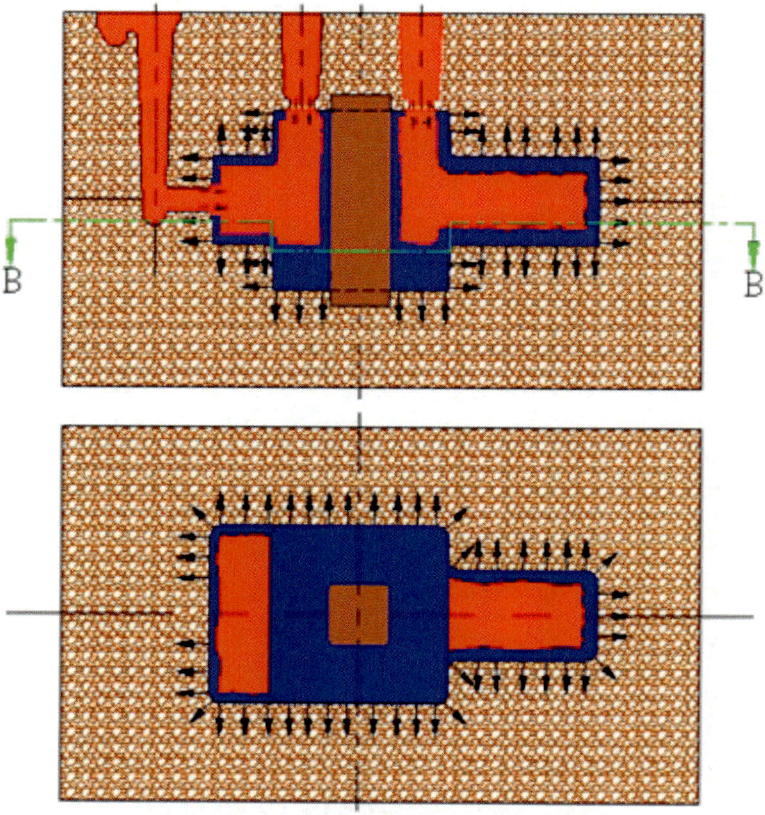

PROGRESSIVE SOLIDIFICATION AT LEVEL B-B

The sectional view at A-A and B-B are example of progressive solidification. The arrow heads marked in drawings are showing direction of flow of heat energy.

The top surface of risers and pouring basin are covered with insulating compound or exothermic compound. The V/A ratio at risers, sprue, and gates are also higher. The cross section of gate is circular. The loss of heat energy from these feeders is less comparison to other areas. The heat energy stored in these feeders will keep the solidification path alive by continuous supply of heat in the solidification path. As such solidification path will turn towards risers and sprue. We can say the solidification is being directed as design.

See the sectional views of directional solidification at level C – C, how the directional solidification has proceeded towards rises and pouring basin in next drawings.

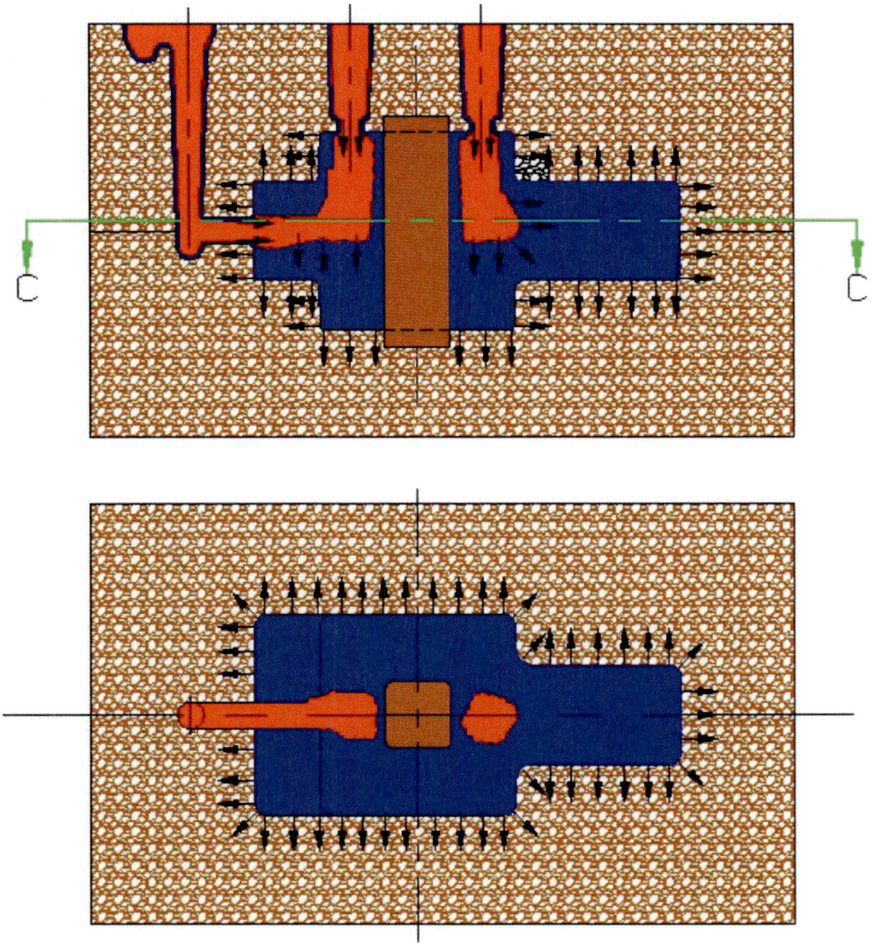

PROGRESSIVE SOLIDIFICATION AT LEVEL C - C

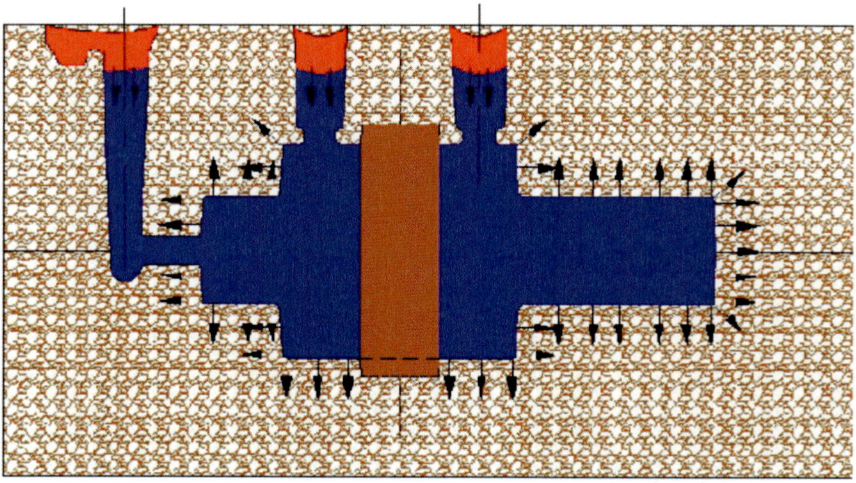

SOLIDIFICATION IN PROGRESS TOWARD POURING BASIN AND RISERS AS DIRECTED

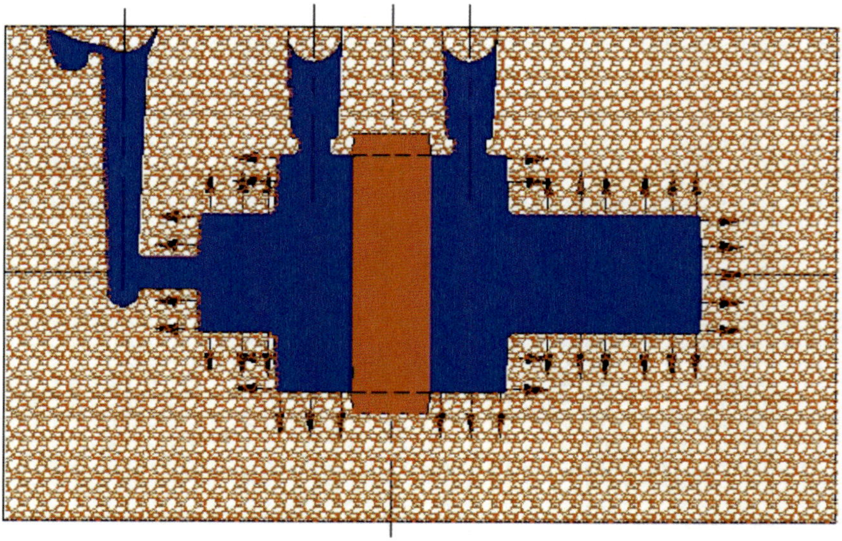

CASTING SOLIDIFIED AS DIRECTED

Chapter-7

Product Design Consideration

Important considerations to have quality cast product:

Consideration of superheat: - Consideration of superheat for a particular composition is very important. It is established on experimental basis.

Insulate risers: – it can be done with insulating sleeves and exothermic compound or insulating compound on top surface of open risers.

Consider V/A ratio: – Here "V" stands for volume of metal and "A" stands for surface area. Higher volume to surface area ratio is always better. It is to be considered while designing gating system and risers.

Heat mass: - Consideration of heat mass is very important for smooth solidification. Never try to feed heavy section through thin section.

Section of casting: – Section of casting to be considered is very important while designing feeding system. Thin section will solidify first; heavy section will help in smooth solidification by feeding liquid metal till it gets solidified.

Geometry of section: - The section of geometry that may create hot spot must be reviewed and considered for redesigning.

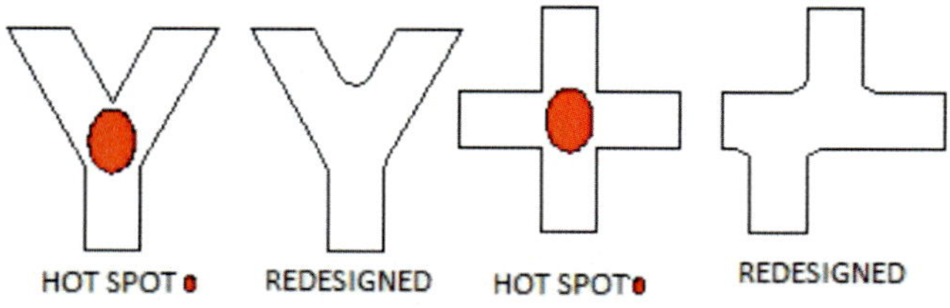

Reduce turbulence: - Design smooth uninterrupted smooth flow path of liquid metal.

Connection between riser and casting: - design short, smooth and having adequate cross section of connecting path.

Taper down sprue: - It is required to reduce the speed of free falling liquid metal to have smooth flow in runner.

Placement of riser: - Locate the riser such that solidification takes place in mould cavity as you desire to have defect free casting.

Permeable core: - Insert permeable core in blind risers where it is required to feed under atmospheric pressure and release gases

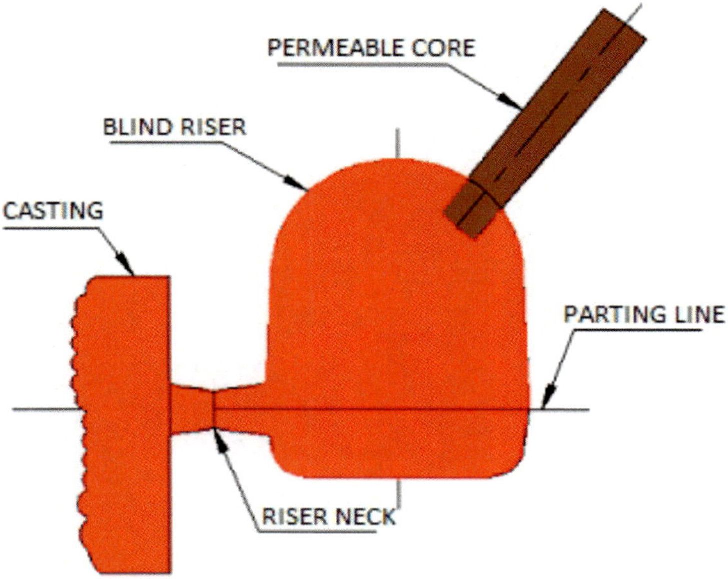

Provide sufficient vent holes in the mould and even in blind risers to provide passage for gasses to escape in the atmosphere.

Provide strainer core at the entry part of sprue to stop dross and impurities from going to mould cavity when it is required.

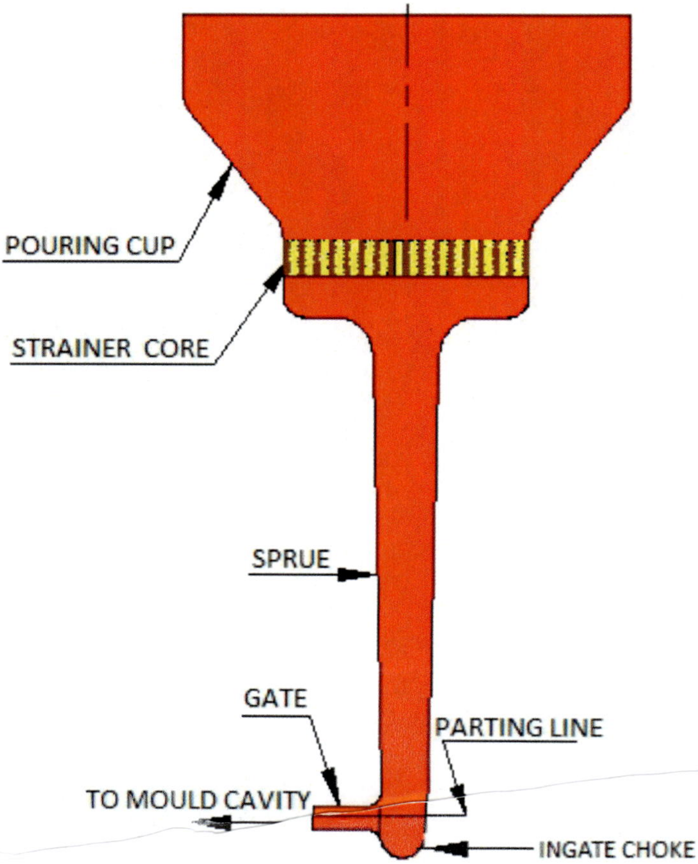

Use chaplets where ever necessary: The chaplets are metallic support used to support core that is not self-supported. It is made with similar material with higher melting temperature than casting material. It assimilates into the cast part when it hardens. The shape and size of chaplet depends upon shape and size of core and its weight.

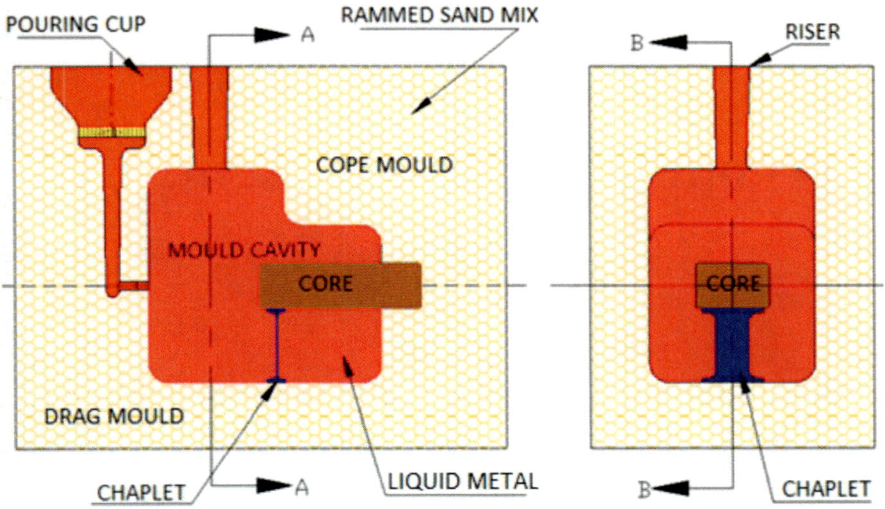

SECTION B - B

SECTION A- A

Chapter-8

Wood, it's Measurement, Use & Seasoning

Wood:

It is natural material abundantly available in forest. The main source is long tall old trees. The cut piece of long tall trunk of tree is called log. The life of tree is assessed by number of annular rings that form every year. The small branches are used as fire wood and the logs are cut into planks of various sizes with the help of power band saw for commercial use. The log contains natural saps and moisture which is driven out by natural and kiln seasoning process.

The wooden plank is measured in length, width and thickness. The unit of measurement is millimeter and inches whereas log is measured in length and circumference. The sale and purchase is being done in cubic feet or cubic meter.

Measurement of wood:

Exercise-1: Find the cost of wood of size 12 ‘x 10’’ x 6’’ @ 1500/= per cubic feet. The 1500 /= is the cost per unit local currency.

The cubic inch or feet is the unit of volumetric measurement. First multiply length x width x thickness, then multiply by cost per unit.

The cost = 12* (10/12) * (6/12) *1500

=12 * 0.8333* 0.5 * 1500 = 5 (Cubic feet) *1500 (Local unit of currency). =7500 (Local unit of currency)

Exercise -2: Find the volume of a log having 3.5 feet periphery and 6 feet long

First find radius of log. It is 2 * 3.142 * R = 3.5

R =3.5 / (2*3.142) =0.55697 (feet)

Volume of log = 3.142 * 0.55697 * 0.55697 * 6 = 5.848186 cubic feet.

Similarly, you have to calculate cost in cubic meter if measurement is in millimetre and meter.

The commercial use of wood:

- Making paper by paper making industry
- Packaging of heavy machines or machine tools
- Making furniture
- Making building
- Manufacturing Commercial ply of various sizes and quality
- Making patterns
- Making models

Suitable wood for Making pattern and core boxes:

The seasoned Teak Wood is regarded as best suitable wood for Pattern making. It has minimum bending tendency as such deformation in size and shape is considerably less than other woods. The mahogany and pine woods are also used.

Seasoning process:

The seasoning is the process of removing natural sap of wood and moisture to minimize deformations in shape and size of structure or manufactured components. The processes of seasoning are:

Natural process:

The cut pieces of wooden logs are kept in the river or pond for 3 to 6 months to wash away the natural sap of wood by the steam of flowing water or dissolved in stagnant water of pond. The logs are taken out from water when sap is washed away and kept in open space to dry out moisture for another 6 months.

The dried log is cut in planks of required thickness and stacked in shed for 3 months again to dry out moisture. The planks should be stacked in such a way that air can circulate horizontally and vertically through the planks.

Kiln seasoning:

It is most common and effective commercial process of wood seasoning. There are two processes - Progressive and Compartmental seasoning.

In the progressive process, the timber loaded trolley is pushed in the chamber from one end. The timer had to pass through differential heating zones from lower to higher and again normal. The natural sap and moistures are dried out while the timer is subjected to various temperature conditions while passing through kiln. The process takes about 2 days to one week depending upon type of wood. The process is quick and used for costly wood.

In the compartmental seasoning the timber remains in the same chamber and subjected to varying temperature condition till sap and moisture is dried up.

Solar Kiln:

In this process the solar energy is used to dry out the natural sap and moisture of wood.

Wood and associated materials for making wooden patterns

Wood: Teak wood, Mahogany, Haldu wood, Pine wood etc. of various length, width and thickness to suit the requirement of individual foundry.

Screws: counter sink headed screws of various sizes to suit the pattern construction requirements.

Hexagonal bolts and nuts of various sizes

Allen bolts and nuts of required sizes

Chapter-9

Wood Working Skills for Making Patterns

General operational knowledge for making pattern with wood:

The pattern maker must be fully conversant with wood working tools, its use and working procedure. The common requirements are

Cutting length, width or thickness: It can be done with hand saw, band saw or various sizes and shape of chisels as and when required for operation.

Planning and making right angle: The surface of wood and side needs marking, cutting and matching with other pieces during the process of pattern construction. It is done with the help of planning machines, hand operated power driven or manual planer. Angle protector or right angle is used to check right angle or bevel edge with respect to surface. The straightness of surface is checked with straight edge. Marking gauge, marking block and scriber are used for marking.

Marking on work pieces: The chalk power is applied on surface to be marked for clear visibility. Construction scale or normal scale, divider, trammel, scriber, marking gauge or marking block is used for it as needed.

Transfer of measurement: It is done from layout or direct from measuring scale as case may be.

Joining wooden planks: The wooden planks are joined together to increase the width or length as needed during construction of pattern. The width is increased by planning side of planks and attaching together with screw, dowel pins and application of adhesive between joining surfaces.

Process of joining wooden planks:

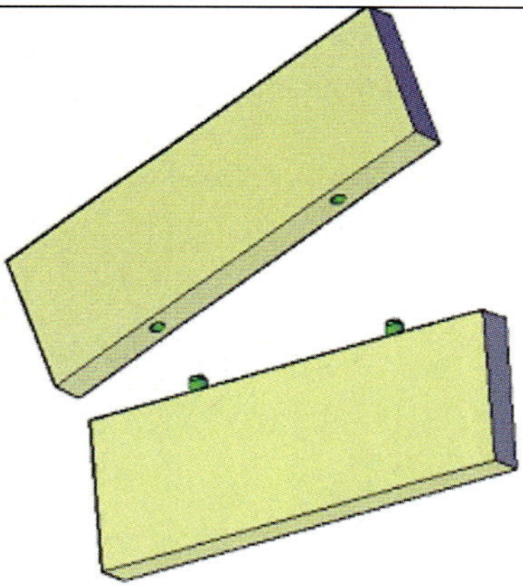	Here one plank has dowel pin and another has matching hole. The wood adhesive is applied in between two planks and dowel pins are fitted by pressing two planks together.
	The two planks are also joined together to increase width by placing screws through pockets in the holes as shown in the drawing. The planks are clamped together by carpenter's clamp for tightening screws.

The width can be increased by joining two pieces with the help of battens placed on one surface of wooden planks. The length is also increased by making suitable joints.

Typical wood working joints:

The wood working joins can be categorised as

- Fixed joints
- Detachable joints

Fixed joints: The specific design and size are decided, made and fixed with wood adhesive, screws and dowel pins as and when needed. These are used

- To increase the length, width or thickness of planks
- In hollow construction of patterns
- In skeleton design of patterns
- In construction of building, furniture or any object that has to remain rigid for prolong period.

Detachable joints: The specific design and size are decided, made and fixed with nut bolts, clamps or any special design to suit the requirements. These are used in construction of

- Foldable furniture
- Mobile wooden equipment such as tent, table, cots etc.
- Core boxes associated with pattern for casting metals
- Sometimes loose pieces of pattern or core boxes

The typical joints as shown below have multiple applications in pattern making based on design.

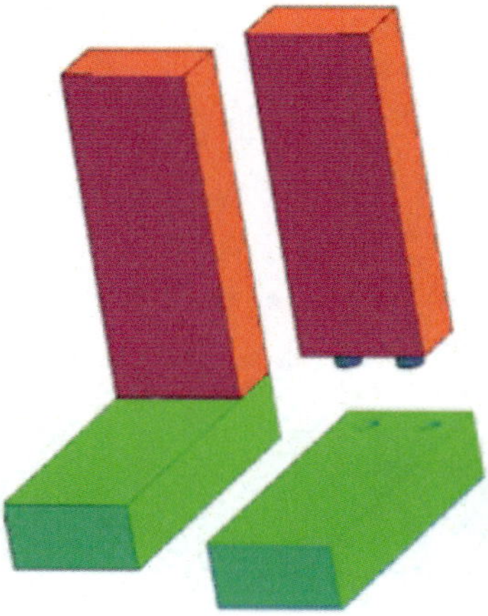

DOWELED BUTT JOINT

RABBET JOINT

LAP JOINT

DOVETAIL JOINT

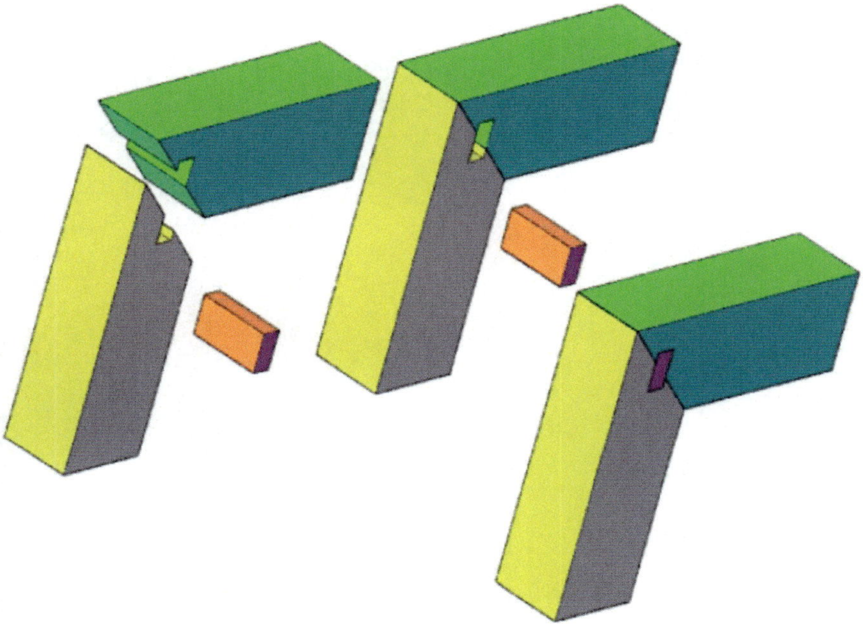

MITER WITH WOOD SPLINE JOINT

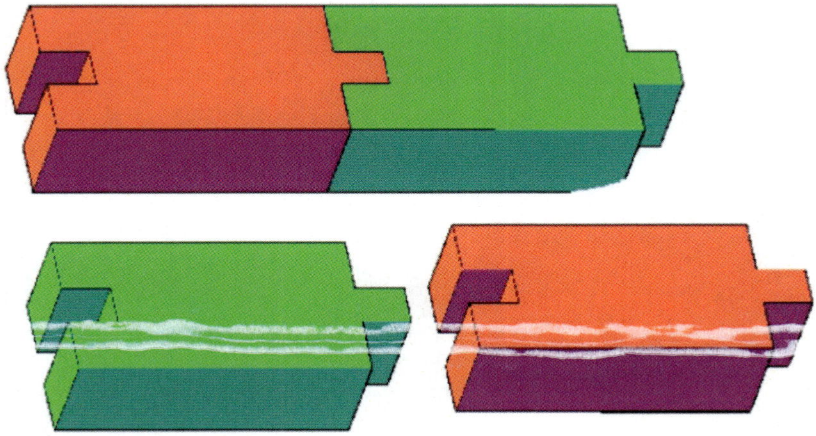

TONGUE AND GROOVE JOINT

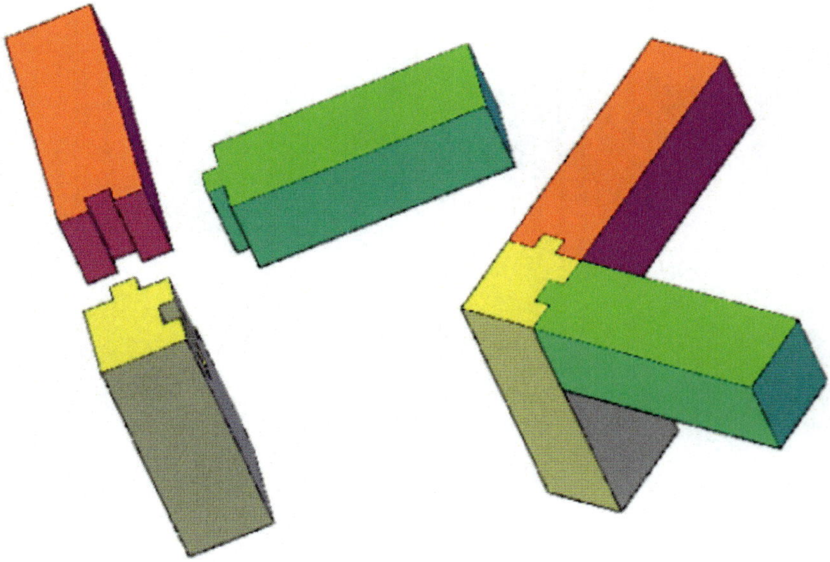

MORTISE AND TENON JOINT

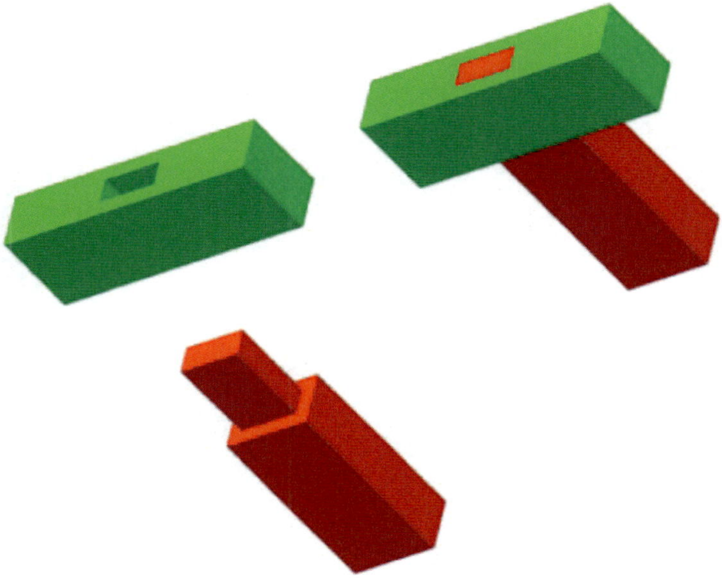

“T’’ JOINT

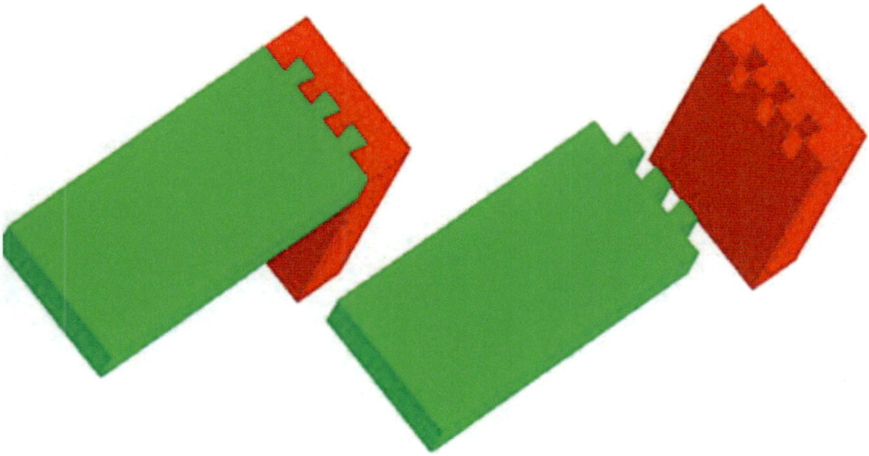

LAPPED DOVETAIL JOINT

HUNCHED STUB TENON CORNER JOINT

HALF LAP JOINT

Chapter-10

Associated Skill for Pattern Making

Associated skills required for pattern making:

- Welding
- Gas cutting
- Soldering
- Brazing
- Welding

Welding

There are various processes of welding, but most common process in use is shield metal arc welding. In this process alternative current or direct current is used from a welding power supply. The consumable electrode coated with flux is used. When electric arc is generated between electrode and metal to be joined, electrode melts and a layer of metal is laid between the pieces to be joined.

The flux coating on electrodes disintegrates giving off vapour. It serves as a shielding gas and provides a layer of slag. It protects the weld area from atmospheric contamination. It is used for welding iron and steel.

The use and selection of electrodes depends upon

- Weld material
- Welding position

Gas cutting:

The gas cutting is also one of the important requirements of pattern making skill. It must be learned for emergency need. Oxy-fuel cutting is the widely used method of common use.

Oxy-fuel cutting:

Acetylene and oxygen gas are used to preheat metal to red hot and then oxygen is used to burn away the preheated metal. The oxidation takes place at the localized place and gap is created in between two ends. It is effective on metals that easily get oxidized at this temperature. Mild steel and low alloy steels being cut easily. It is suitable to cut 1/4 inch to 12 inch thick.

When acetylene mixed with oxygen is ignited the flame temperate rises between 5700° F to 6300°F. The acetylene is highly combustible when mixed with oxygen.

It can cut low to medium carbon steel and wrought iron but it can't cut high carbon steel, cast iron and stainless steel

Apparatus used:

- Regulator
- Gas hoses
- Non return valve
- Check valve
- Torch

Type of torch used:

- Welding torch
- Cutting torch
- Rose bud torch
- Injector torch

Non return valve:

It is fitted between hose and torch on both oxygen and fuel (acetylene) lines to prevent flame of oxygen fuel mixture being pushed back. Pushing back into either cylinder will damage the equipment or cylinder may explode.

Check valve:

It allows gas to flow in one direction only.

Welding torch:

It is used to weld metal

Cutting torch:

It is used to cut material.

Safety:

Welding screen with blue glass must be used to protect eyes.

Safety of cylinder is also very important; one must check possibility of any leakage or explosion.

Keeping safe against any chemical exposure is also important.

Must be alert against any flash back while welding or gas cutting.

Soldering:

The soldering is the process of joining two or more metal items together by melting and flowing filler metal (called solder) into the joints. The filler metal used has lower melting point than the adjoining metal.

Brazing:

It is metal joining process where in filler metal is heated above melting point and distributed between two or more close fitting parts. The torch brazing is most common method of mechanized brazing in use.

Bench work:

The bench work is associated with the skill of cutting metal, filing, drilling and tapping. You have to learn metal cutting with hack saw by hand or with power driven hacksaw or band saw. The filling work is done with the help of hand files of various shape and size.

The drilling work is carried out with fixed drill machine where tapping work is required

Selecting required size of drill for particular size of tap is very important.

In case of metric thread, it is taken as TD = MD – PITCH Where TD = tap drill size, MD = Major diameter of the tap

Suppose you want to make a tap hole of M10 having 1.5 pitches, then diameter of drill will be 10 -1.5 = 8.5 mm.

Chapter-11

Machine and Tools for Pattern Making

Machines required by wooden Pattern shop:

Wood planning machines

- Surface planner
- Thickness planer

Drilling machine

- Plate shearing machine
- Plate bending facilities (Bending machine or dies for manual bending)
- Wood cutting bandsaw
- Bandsaw blade jointer
- Disc sander
- Spindle sander
- Spindle tool grinder

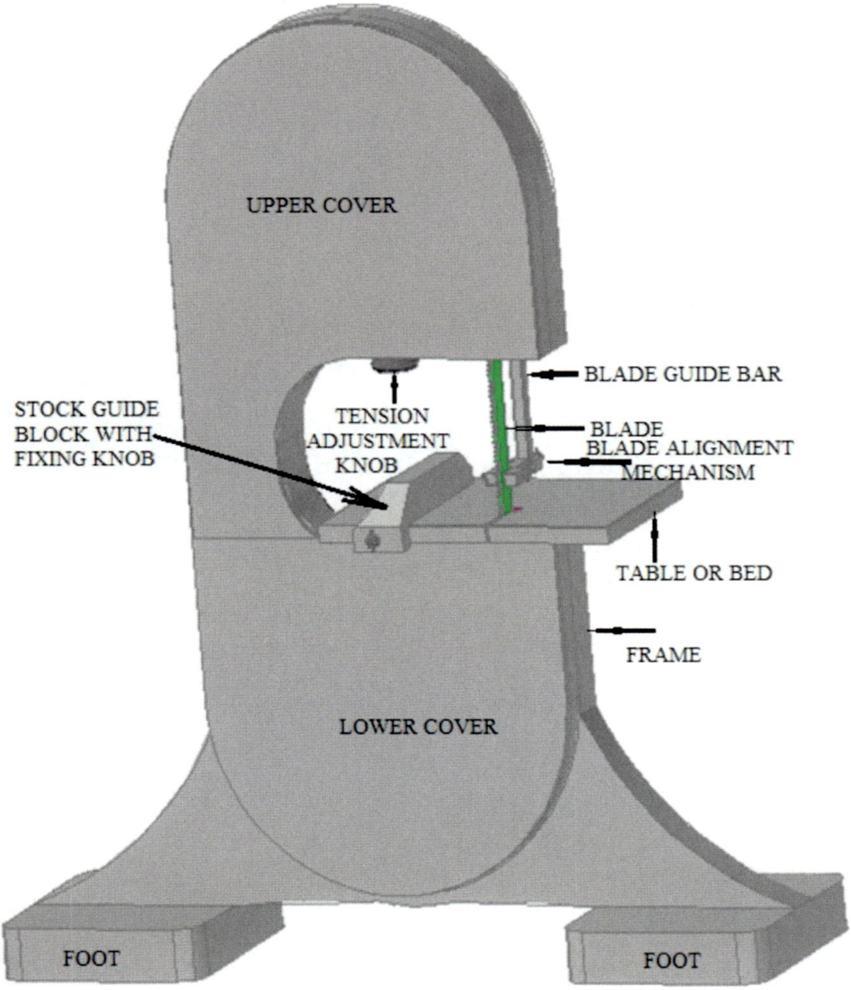

- Metal cutting bandsaw
- Bandsaw blade grinder

Hand tools:

- Power driven portable planer
- Hand planer

- Wood cutting tools

Right Angle

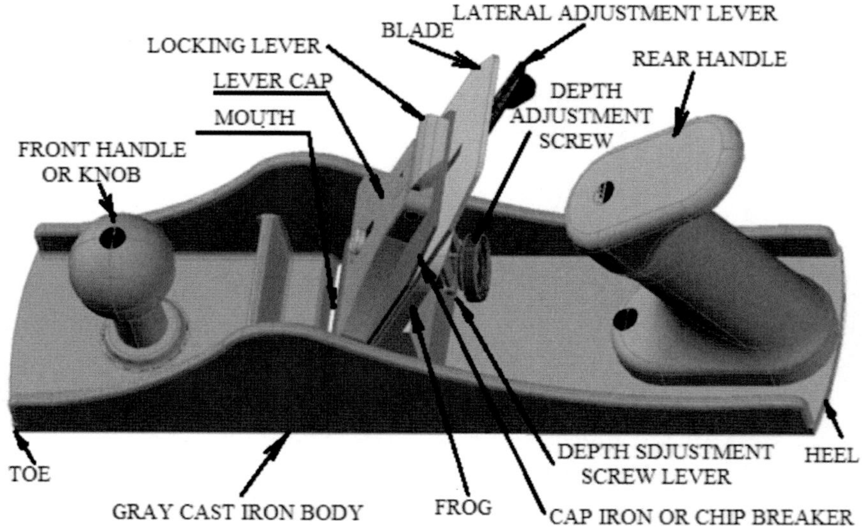

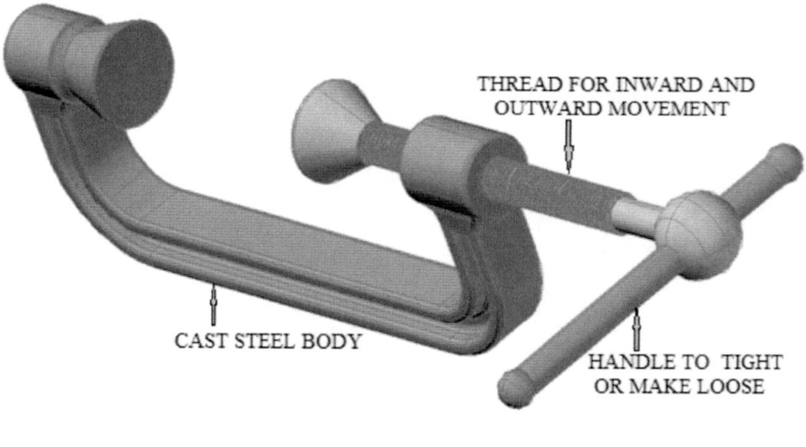

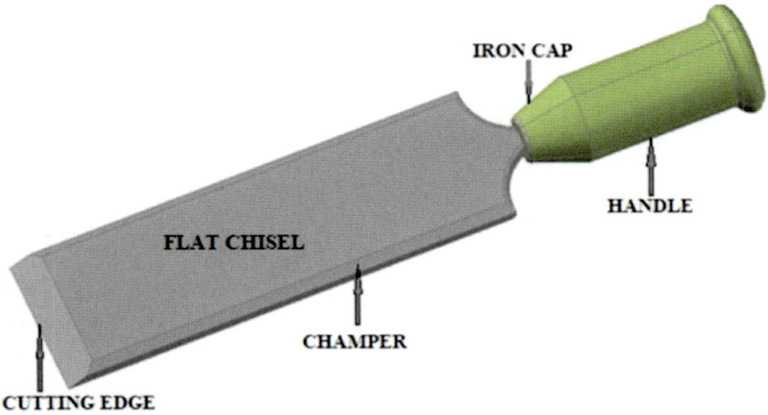
IRON CAP
HANDLE
FLAT CHISEL
CHAMPER
CUTTING EDGE

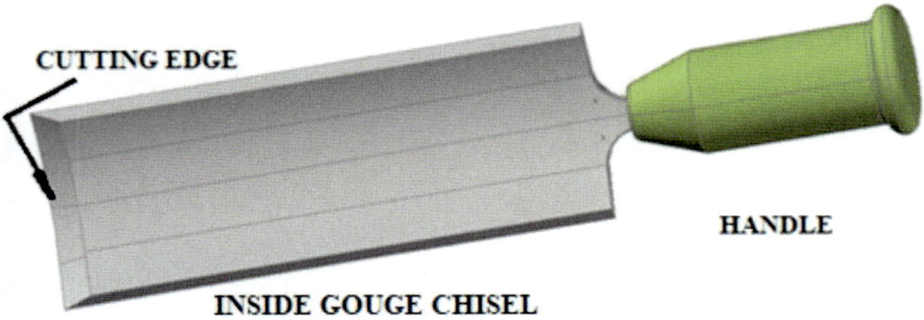
CUTTING EDGE
HANDLE
INSIDE GOUGE CHISEL

HANDLE
OUTSIDE GOUGE CHISEL
CUTTING EDGE

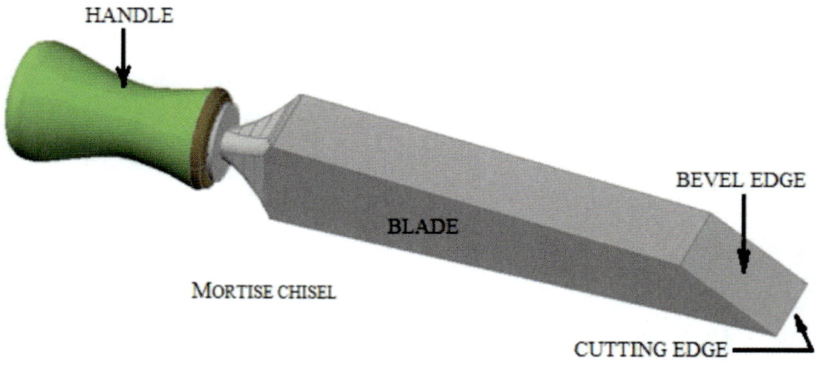

- Spanners and pliers
- Drill bits
- Bench grinder
- Disc sander
- Callipers (inside and outside)
- Bevel protectors
- Auger bits
- Compass
- Beam compass
- Gauge or Round chisel for cutting round profile
- Screw driver (power driven or manual)
- Spanners
- Ratchet
- Hammer
- Mallet
- Handsaw
- Files
- Vernier calliper
- Divider
- Odd leg calliper
- Marker (scriber)
- Marking gauge

- Lay-out Board

Chapter-12

Inspection of Pattern

Inspection of patterns

It is carried out to stop supply of defective pattern and core boxes to moulding shop for production. The probable defects (mistakes) in pattern are

- Poor surface finish
- Wrong selection of parting line or parting line has not been properly aligned with entire body of pattern: – it may create problem during withdrawal from mould cavity after ramming sand.
- Poor strength: – the construction process may be defective, thin section of wood might have been used. The pattern with poor strength can't be used for repetitive use as will not have sufficient strength to bear the moulding impact.
- Miss match in mounting on match plate or mounting on cope and drag board.
- Draft is inadequate.
- Draft may be missing.
- Missing fillet or radius.

- Mistake in lay out
- Dimensional mistake in layout
- Measurement of pattern size not matching with layout.
- Mistake in understanding component drawing.
- Machining allowance missing
- Wrong placement of gates and risers.

Inspection format:

Description	Compliance	Remarks
Lay-out		
Parting line		
Surface finish		
Strength		
Fillets		
Radius		
Draft		
Machining allowance		
Position of gate and riser		
Sizes		

Chapter-13

Mould and core making

Mould making:

It is the process of forming internal cavity in the moulding flask with moulding sand having entry path for liquid metal and feeders to control solidification to get defect free castings. The cavities are formed with the help of pattern which is called mould.

Mould:

It is the negative shape of pattern formed in moulding flask by ramming special sand mix all around the pattern surfaces in the foundry. The components of mould are pouring basin, sprue, runners, ingates, sprue well, ingate choke and risers. It is used for producing casting by pouring molten metal.

Moulding Process:

The following steps are involved in the preparation of mould:

- Making pattern to suit the requirement of customer
- Setting moulding flasks over the pattern
- Applying releasing agent all over surfaces of pattern, moulding board and gating system arrangement such as risers, runners, sprue, ingates etc.
- Ramming sand mixture all over the pattern surfaces to have drag and cope parts
- Hardening of sand mix or allowing it to get harden and develop its bond while in contact with pattern
- Withdrawal of pattern from mould cavity.
- Applying ceramic coat after repairing or mending fillets, radius and cleaning any loose sand
- Assembly of cope and drag

- Clamping cope with the drag rigidly to prevent it from floating when metal is poured.

Mould Making Processes

Hand moulding:

It is used in small foundry. Wooden rammer or pneumatic rammer is used by moulders to pack sand mixture on all over surfaces of pattern. The sand mixture is retained by flask that provides its outer shape.

Mechanised moulding process:

In this process the sand mixture is packed all over surfaces of pattern, risers and getting system attached to pattern by sand slinger, jolting or squeeze devices.

Automatic high pressure sand moulding:

In this process automatic jolting and vibrators are used to prepack sand mixture all over pattern surfaces in moulding flask and finally compressed air pressure power piston used to compact the mould.

Horizontal sand flask moulding:

In this process sand mixture is shot or slung down in a moulding flask all over pattern surfaces and finally squeezed with hydraulic pressure.

Match plate moulding:

In this process, the cope and drag parts of patterns are mounted on the opposite sides of a wooden board or steel plate called match plate. The moulding flasks are first set on both sides of match plate over pattern. The moulding sand is slingered in boxes and boxes are finally squeezed to compress the mixture.

The flasks are to be pulled out in opposite direction and pattern with match plate is separated from flasks. The separated flasks will be the cope and drag moulds. The entire operation is done by automatic machine.

Moulding material:

The sand mixture is the main material used for green sand casting. The common composition is-

70 – 85 % Silica Sand

10 – 12 % bonding materials: Clay and cereals etc.

3 – 6 % Water.

The composition of sand mixture is variable and determined by Research and Development Laboratory on the basis of experiments conducted to suit their requirement considering variables such as method of moulding, metal composition, quality, metal transfer and pouring system.

The quality of sand mixture plays a greater role in having good quality cast as such it should fulfil the following characteristics.

Strength:

It is the ability of sand to maintain mould cavity shape while molten metal is poured and retain strength throughout the solidification period.

Permeability:

It is the ability of sand mix to pass out trapped gases, air and water vapour through it from mould cavity. It is determined by the size and the shape of the sand grains. A higher permeability can increase the porosity of the mould and may promote metal penetration, but a lower permeability will produce better surface finish.

Flowability: The flowability is the ability of moulding sand mixture to move like as fluid to each and every corner and surface of pattern and accessories.

Thermal stability:

It is the ability to resist damage and cracks formation on surfaces, corners and in any part of mould cavity due to heat of molten metal.

Collapsibility:

It is the ability of sand to collapse or compress during solidification of casting. In lack of compressibility casting will not be able to shrink freely in the mould and cracks may develop.

Cohesiveness:

It is the ability of sand grain to stick and bond together when packed over pattern in the flask

Reusability:

It is the ability of sand to be used again for future moulding.

Core:

It is body of proper sand mix made with the core box by appropriate core making process to suit individual foundry men's requirement. When it's placed in the mould cavity, it blocks the passage of molten metal but allows it to pass all over surface except core print area that fits in its seat. Its placement in mould cavity generally provides internal shape in casting but sometimes used to form external shape.

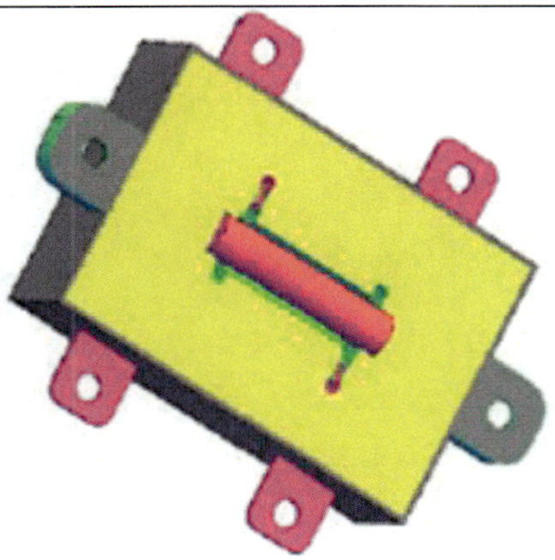	The red coloured component is core that has been set in cavity made with core print. The green visible area is mould cavity.

Core making process:

Open the split box or dismantle the assembly of loose pieces if it is made with loose pieces.

Apply releasing agent all over surface of cavity in the core box made for it that will come in contact with sand mixture, and reassemble it.

Bring pre-prepared core sand mix or make sand mixture to suit the need of your casting.

Ram the core sand mixture in core box by wooden rammer, pneumatic rammer, jolting machine, squeezing or blowing machine as needed or with available facilities.

If needed provide reinforcement in the body of sand before ramming. The reinforcement material may be wire or nail. The size of nails and wire should provide strength but must be inside the sand mixture.

Let the core remain in the box to get bonding strength if it is air setting sand mix, blow carbon dioxide if binder is sodium silicate or put it in the oven after removing from core box if backing is required to get bond and strength.

Releasing agent / parting compound:

It is used for easy removal of pattern from mould cavity or core from core box. The common powder in use is talc, graphite and dry silica. The common liquid in use is mineral oil and water based silicon solutions.

Core Sand:

The sand mixture is made by foundry men to suit their requirement of casting considering metal composition, casting process and available facilities. The quality of binder used for core sand is very important.

Binders used in sand mixture for core:

Sodium silicate: It is most common binder in use. It gets hardened and sand grains bind together providing sufficient strength to core when carbon dioxide is blown through the packed mixture in core box. The other binders in use are

- Silica gel
- Oil binder
- Sulphite
- Starch

- Cereals
- Molasses

Resin: Resin binders are synthetic binder. Commonly used resin binders are urea formaldehyde and phenol formaldehyde. These binders have some advantages over others such as good collapsibility, low gassing and good surface finish.

Properties of good binder

A good binder will

- Maintain Strength
- Provide collapsibility: It is very important for core. It is the property to hold the sand grains together as long as metal solidifies in mould and then burnt out losing its bonding power.
- Provide ability to maintain shape and will not allow distorting core during baking in oven.
- Maintain strength during storage before and after baking.
- Absorb no or minimum moisture during storage
- Maintain flowability in mixture to flow every corner of core box during core making process

Shell core:

It is a hollow body of special sand mixture whose outer shape is formed with thin layers. The outer body has same shape and size to form inner profile of component. It is used for internal shape of casting where machining is not possible. The example of such shape is cylinder block and cylinder head of petrol engine cars. To cast cylinder block and cylinder head shell cores are assembled together matching with each other. The assembly of these shell cores provide cavity to have outer shape and core to form inner profiles. The assembly of cores serves the purpose of mould to provide the outer body with molten metal and self-core made in it inner profile of component. The castings made with these cores are having good finish on outer as well as inner surface of castings, but sometimes machining operation is carried out on matching surfaces.

Shell core making process:

The entire core making process is done by automatic processing machine.

- The split core box is opened and releasing agent is applied on inside cavity that will form core.
- Both parts are assembled and heated up to required temperature.
- The pre mixed Zirconia sand mixture with epoxy resin and phenol formaldehyde binder is blown under pressure on the walls of core cavity. The moment mixture touches the hot surface of cavity reaction take place and a thin layer of mixture is formed. The shape of core looks like as solid body.
- The core is taken out when core box opens at parting line

Chapter-14

Sand Blasting & Casting Defects

Sand Blasting and fettling:

It is the process of removing fins, risers and gating arrangements that are attached to casing during moulding process for feeding mould cavity. The surface of casting is cleaned by sand blasting process. The various processes are used in work.

Probable defects in castings:

- Gas defects
- Shrinkage cavities
- Moulding material defects
- Pouring metal defects
- Metallurgical defects.

Gas defects:

It is caused by entrapped gas due to poor permeability and venting system

- Blow holes
- Scar
- Blister
- Gas holes

- Pin holes

Shrinkage cavity:

It is due to shrinkage in the casting during solidification process. It is due to improper feeding system. The shape of shrinkage cavity may be like as funnel.

Moulding material defects:

It is caused due to

- Insufficient strength in moulding sand mixture.
- Improper ramming
- High velocity of molten metal can be reduced by reducing turbulence

Pouring metal defects:

It is caused by higher super heat and pouring rate. Metal penetration is caused by it.

Metallurgical defect:

- Hot tear
- Hard spot

Misran:

Casting solidifies before mould cavity is filled

Cold shut:

Fusion do not take place in two portion of cast part due to low pouring temperature, slow pouring or thin cross sections.

Hot tears and cracks:

It is caused by low mould collapsibility. They occur when metal is restrained from contraction during solidification. The proper mould design can solve this problem.

Mismatch:

It is due to shift in cope and drag mould and mistake in mounting of cope and drag patterns

Chapter-15

Q&A

1. What is the use of pattern and why knowledge of pattern making is essential?
2. What is core print?
3. What is the difference between internal and external core?
4. Differentiate between shrinkage and contraction allowance.
5. What is the importance of parting line?
6. What is draft?
7. What is machining allowance?
8. What is the difference between fillet and radius?
9. What is the difference between shake and distortion allowance?
10. Name the categories of patterns on the basis of construction material.
11. Name the category of wooden patterns on the basis of shape, size and required quantity of castings.
12. What is master pattern? Why it is essential?
13. What is metal pattern and why it is made?
14. When thermocol is used for making pattern?
15. What is follow board pattern?
16. What is the difference between solid pattern and split pattern?
17. What is self-core pattern?
18. What are the elements of gating system?
19. Why fillet is important at the junction of the sprue and runner?

20. What is riser?
21. What are the advantages of top riser?
22. What are the advantages of blind riser
23. What is V/A Ratio?
24. What is progressive solidification?
25. What is directional solidification?
26. What is permeable core?
27. What is the direction of solidification in top gating system?
28. What is turbulence and what it has bad effect?
29. What is mould?
30. What is core?
31. What is the importance of releasing compound in making mould?
32. What are the probable casting defects in casting?
33. What are the properties of good binder?
34. What is shell core?
35. What is seasoning of wood and why is it essential?

Answers to Questions in Chapter 15

1. The pattern is used for casting objects that are not possible to make by other manufacturing processes. The pattern making is very important in foundry technology.

2. See description on chapter -1

3. The internal core is made to have internal cavity or shape of the casting. It also has shape and size of core print to set in the core print cavity associated with mould. The external core is made to cast the shape which is simply not possible to make with pattern. The cylinder block and cylinder head castings are made with shell cores.

4. The volumetric reduction in shape and size of total cast metal poured in mould cavity from red hot liquid to red hot solid stage is called shrinkage. The additional volume of metal in addition to cast part, sprue, runner and ingrate is called shrinkage allowance. It is compensated by providing appropriate volume of riser in proper location.

   The red-hot solid metal again has volumetric reduction while cooling down from red-hot solid stage to room temperature. This volumetric reduction is compensated by addition of linear measurement on all sizes in the drawing of object to be cast while making layout. The contraction allowance is added according to contraction behavior determined on experimental basis for the metal to be cast.

5. It is the separation line or plane that divide split pattern in two parts but in case of solid pattern it is separation plane of cope and drag parts of mould. It is very important reference line for providing draft.

6. It is the taper or inclination provided on all vertical surfaces of pattern with reference to parting surface. It facilitates

easy withdrawal of pattern after ramming sand mixture all-round the surfaces.

7. The machining allowance is the additional material added on surfaces that needs machining. The quantity or thickness of additional material depends upon requirement of surface finish.

8. The rounding of internal corner of pattern (Intersection of inclined or vertical plane and horizontal) is termed as fillet whereas rounding of external corner (intersection of horizontal plane and vertical or inclined plane) is termed as radius.

9. The shake allowance is the negative allowance. The pattern is increased by size to take care of wrapping force applied all-round the surfaces to facilitate easy removal of pattern. There is no specific measure. It is provided on approximation basis.

   There is no increase or decrease in the size of pattern. It is applicable for thinner section of casting. The pattern is distorted in opposite direction on the basis of expectation and design of gating system.

10. See Chapter – 3

11. See Chapter – 4

12 The master pattern is made of wood. It is required to make metal pattern. The metal pattern can't be made directly with metals. The master pattern is made with addition of extra

metal on all over surfaces of master pattern to get machined surfaces having machining allowances.

13 The metal pattern is also pattern equipped with allowances, risers, runners and gating system. It is made with aluminum alloys, cast iron and steel. It is made for mass production of repetitive castings.

14 See chapter – 3

15 See chapter – 4

16 The solid patterns are generally made for casting simple shape that can be easily withdrawn from mould cavity. It has parting line or surface at one end face.

Split patterns are made in two parts joined together with dowel pins. The shapes are not easy to cast with single piece solid pattern. The parting plane is selected at suitable location for easy withdrawal from cope and drag parts of mould cavity

17 When the component or part has a cavity that can be used as core box for making core without changing the shape of the casting in pattern making is called self-core pattern.

18 See Chapter – 6

19 It is essential to reduce possibility of washing out of sand mixture from mould cavity and eliminate inclusion of sand in the casting.

20 The riser is a feeder or reservoir that feeds the metal to casing during solidification shrinkage from red hot liquid stage to red hot solid stage.

21 See Chapter – 6
22 See Chapter – 6
23 See Chapter – 6
24 See Chapter – 6

25 The directional solidification is the process of diverting the solidification path in our planned way by proper design of risers, runners, ingates, sprue and pouring basin.

26 The permeable core is a special shape core made with special sand mixture to have sufficient permeability to scape gas from mould cavity after pouring molten metal. It is inserted in blind riser having top surface exposed on top surface of cope mould.

27 The solidification of metal in top gating system starts from bottom of the casting.

28 See Chapter - 6
29 See Chapter - 13
30 See Chapter - 13
31 See Chapter - 13
32 See Chapter - 14
33 See Chapter - 13

34 The shell core is the special purpose core made with premixed zirconia sand mixture with epoxy resin and phenol formaldehyde binder. It has outer shape same as the internal

profile of casting with additional associated shape to fit in assembly. It has hollow body with uniform wall thickness.

35 The seasoning is the process of removing natural sap of wood and moisture from the logs and cut planks. It is essential to minimize deformation or warpage in manufactured goods such as Patterns, core boxes, furniture and house hold goods.

Attributions

Image of Iron Foundry in Zhenhai by Jim Bowen / CC BY 2.0 has been used to create the book cover

Cover Photo Link:

https://bit.ly/2Udx5Na

Author of the Cover Photo: Jim Bowen

https://www.flickr.com/photos/jamiedfw/

Attribution of the Image is under Attribution 2.0 Generic **(CC BY 2.0)**

I would love to hear feedback and inputs from readers and practitioners. You can provide your inputs and feedback by emailing at reesaa@indiavivid.com or visiting www.reesaa.com

- Sheojee Prasad

Made in United States
Orlando, FL
04 November 2026

90208350R00076